Genetics and Genomics of Plants

Edited by Arit Efretuei

Edition 2018

ArclerPress

Published by

Arcler Press LLC

708 3rd Avenue, 6th Floor

New York

NY 10017

United States of America

www.arclerpress.com

ISBN: 978-1-68094-558-4 (Hardback)

Printed in United States of America

Library of Congress Control Number: 2017933129

GENETICS AND GENOMICS OF PLANTS

Edited by
Arit Efretuei

ArclerPress

About the Editor

Arit Efretuei

Arit Efretuei obtained her PhD in Agriculture from the University of Reading, UK in 2015. Her interests are in crop nutrient uptake, soil nutrient management and mitigation of soil nutrient losses to the environment. She is presently working as a freelance agronomic writer.

Table of Contents

3 GENOMICS AND PLANT GENOMES

4 DOUBLED HAPLOIDY

5 GENETIC ENGINEERING AND PLANTS

6 GENETICALLY MODIFIED TREE

7 GENETICALLY MODIFIED CROPS

8 CHLOROPLAST CAPTURE

List of Abbreviations

AGS	Advanced Genetic Sciences
AP/MS	Affinity Purification and Mass Spectrometry
CATH	Class, Architecture, Topology and Homologous superfamily
dNTPs	Deoxynucleosidetriphosphates
DNA	Deoxyribonucleic Acid
DNMTs	DNA Methyltransferases
DH	Doubled Haploid
DL	Dynamic Loop
EPA	Environmental Protection Agency
EASAC	European Academies Science Advisory Council
FDA	Food and Drug Administration
GMO	Genetically Modified Organism
GBRI	Geneva Biomedical Research Institute
HCCB	Hydrocarbonoclastic Bacteria
IRRI	International Rice Research Institute
JCSG	Joint Center for Structural Genomics
LADs	Lamina-Associated Domains
NIH	National Institute of Health
OSTP	Office of Science and Technology
ORF	Open Reading Frame
OECD	Organization for Economic Co-operation and Development
PCBs	Polychlorinated Biphenyls

PSI	Protein Structure Initiative
QTL	Quantitative Trait Loci
USD	US Dollars

Preface

Plant genetics and genomics is a growing and constantly evolving field of study. Plant genetics deals with heredity in plants, specifically mechanisms of hereditary transmission and variation of inherited characteristics. Plant genetics differs from animal genetics in a number of ways: somatic mutations can contribute to the germ line more easily as flowers develop at the end of branches composed of somatic cells; polyploidy is more common; and plants additionally contain chloroplastic DNA. Like mitochondria, chloroplasts have their own DNA, complicating pedigrees somewhat. Like animals, plants have somatic mutations regularly, but these mutations can contribute to the germ line with ease, since flowers develop at the ends of branches composed of somatic cells. People have known of this for centuries, and mutant branches are called "sports". If the fruit on the sport is economically desirable, a new cultivar may be obtained. Some plant species are capable of self-fertilization, and some are nearly exclusively self-fertilizers. This means that a plant can be both mother and father to its offspring, a rare occurrence in animals. Scientists and hobbyists attempting to make crosses between different plants must take special measures to prevent the plants from self-fertilizing. Plants are generally more capable of surviving, and indeed flourishing, as polyploids. Polyploidy, the presence of extra sets of chromosomes, is not usually compatible with life in animals. In plants, polyploid individuals are created frequently by a variety of processes, and once created usually cannot cross back to the parental type. Polyploid individuals, if capable of self-fertilizing, can give rise to a new genetically distinct lineage, which can be the start of a new species. This is often called "instant speciation". Polyploids generally have larger fruit, an economically desirable trait, and many human food crops, including wheat, maize, potatoes, peanuts, strawberries and tobacco, are either accidentally or deliberately created polyploids. The use of genetically engineered crops has helped many farmers deal with pest problems that reduce their crop production. The impact of pest-resistant crops has led to a much higher yield for farmers in today's world. They can use less pesticides which reduces the chemicals that they put into the ground. Certain engineered crops have led to farmers all over the world to increase crop yield exponentially in recent years.

Content Coverage

Chapter One presents an overview of plant genetics that deals with heredity in plants, specifically mechanisms of hereditary transmission and variation of inherited characteristics.

Chapter Two focuses on the chromatin, centromere, and telomeres. Chromatin is a mass of genetic material composed of DNA and proteins that condenses to form chromosomes during eukaryotic cell division.

Chapter Three focuses on genomics and plant genomes. Genomics aims at the collective characterization and quantification of genes, which direct the production of proteins with the assistance of enzymes and messenger molecules.

Chapter Four focuses on doubled haploid (DH), a genotype formed when haploid cells undergo chromosome doubling. Artificial production of doubled haploids is important in plant breeding.

Chapter Five is concerned with genetic engineering in plants. Genetically engineered organisms currently on the market include plants with resistance to some insects, plants that can tolerate herbicides, and crops with modified oil content.

Chapter Six focuses on the concepts of tree breeding and the application of genetic, reproductive biology and economics principles to the genetic improvement and management of forest trees.

Chapter Seven focuses on the genetically modified crops (GMCs, GM crops, or biotech crops) used in agriculture, the DNA of which has been modified using genetic engineering techniques. The aim is to introduce a new trait to the plant which does not occur naturally in the species.

Chapter Eight presents the study of chloroplast capture, an evolutionary phenomenon that an inter-species hybridization and subsequent backcrossing genetic exchanges would yield a plant with new genetic combination.

CHAPTER 1

Plant Genetics

Chapter Contents

- Introduction
- Features of Plant Biology
- DNA
- Gregor Mendel
- Genetically Engineered Crops

Plant genetics deals with heredity in plants, specifically mechanisms of hereditary transmission and variation of inherited characteristics. Plant genetics differs from animal genetics in a number of ways: somatic mutations can contribute to the germ line more easily as flowers develop at the end of branches composed of somatic cells; polyploidy is more common; and plants additionally contain chloroplastic DNA.

INTRODUCTION

Plant genetics is a very broad term. There are many facets of genetics in general, and of course there are many facets to plants. The definition of genetics is the branch of biology that deals with heredity, especially the mechanisms of hereditary transmission and the variation of inherited characteristics among similar or related organisms. And the definition of a plant is any of various photosynthetic, eukaryotic, multicellular organisms of the kingdom Plantae characteristically producing embryos, containing chloroplasts, having cell walls which contain cellulose, and lacking the power of locomotion. Although there has been a revolution in the biological sciences in the past twenty years, there is still a great deal that remains to be discovered. The completion of the sequencing of the genomes of rice and some agriculturally and scientifically important plants (for example Physcomitrella patens) has increased the possibilities of plant genetic research immeasurably.

An image of multiple chromosomes, making up a genome

FEATURES OF PLANT BIOLOGY

Plant genetics is different from that of animals in a few ways. Like mitochondria, chloroplasts have their own DNA, complicating pedigrees somewhat. Like animals, plants have somatic mutations regularly, but these mutations can contribute to the germ line with ease, since flowers develop at the ends of branches composed of somatic cells. People have known of this for centuries, and mutant branches are called "sports". If the fruit on the sport is economically desirable, a new cultivar may be obtained.

Some plant species are capable of self-fertilization, and some are nearly exclusively self-fertilizers. This means that a plant can be both mother and father to its offspring, a rare occurrence in animals. Scientists and hobbyists attempting to make crosses between different plants must take special measures to prevent the plants from self-fertilizing.

Plants are generally more capable of surviving, and indeed flourishing, as polyploids. Polyploidy, the presence of extra sets of chromosomes, is not usually compatible with life in animals. In plants, polyploid individuals are created frequently by a variety of processes, and once created usually cannot cross back to the parental type. Polyploid individuals, if capable of self-fertilizing, can give rise to a new genetically distinct lineage, which can be the start of a new species. This is often called "instant speciation". Polyploids generally have larger fruit, an economically desirable trait, and many human food crops, including wheat, maize, potatoes, peanuts, strawberries and tobacco, are either accidentally or deliberately created polyploids.

Hybrids between plant species are easy to create by hand-pollination, and may be more successful on average than hybrids between animal species. Often tens of thousands of offspring from a single cross are raised and tested to obtain a single individual with desired characteristics. People create hybrids for economic and aesthetic reasons, especially with orchids.

DNA

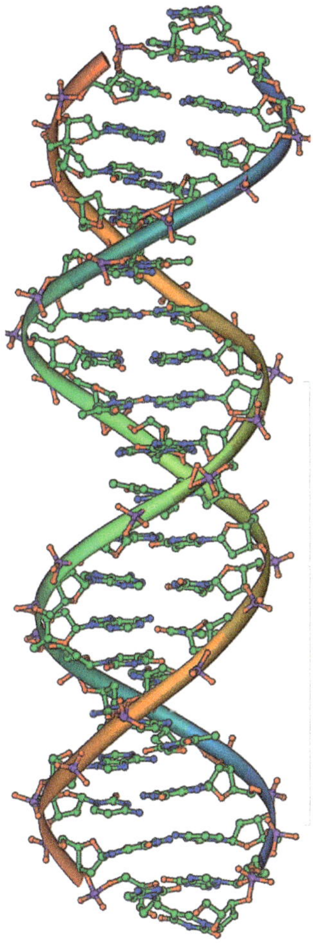

The structure of part of a DNA double helix

Deoxyribonucleic acid (DNA) is a nucleic acid that contains the genetic instructions used in the development and functioning of all known living organisms and some viruses. The main role of DNA molecules is the long-term storage of information. DNA is often compared to a set of blueprints or a recipe, or a code, since it contains the instructions needed to construct other components of cells, such as proteins and RNA molecules. The DNA segments that carry this genetic information are called genes, but other DNA sequences have structural purposes, or are involved in regulating the use of this genetic information. Geneticists, including plant geneticists, use this sequencing of DNA to their advantage as they splice and delete certain genes and regions of the DNA molecule to produce a different or desired genotype and thus, also producing a different phenotype.

GREGOR MENDEL

Gregor Mendel was an Augustinian priest and scientist born on 20 July 1822 in Austria-Hungary and is well known for discovering genetics. He went to the Abbey of St. Thomas in Brno. He is often called the father of genetics for his study of the inheritance of certain traits in pea plants. Mendel showed that the inheritance of these traits follows particular laws, which were later named after him. The significance of Mendel's work was not recognized until the turn of the 20th century. Its rediscovery prompted the foundation of the discipline of genetics allows geneticists today to accurately predict the outcome of such crosses and in determining the phenotypical effects of the crosses. He died on 6 January 1884 from chronic nephritis.

Modern Ways to Genetically Modify Plants

There are two predominant procedures of transforming genes in organisms: the "Gene gun" method and the *Agrobacterium* method.

"GENE Gun" Method

The "Gene Gun" method is also referred to as "biolistics" (ballistics using biological components). This technique is used for in vivo (within a living organism) transformation and has been especially useful in transforming monocot species like corn and rice. This approach literally shoots genes into plant cells and plant cell chloroplasts. DNA is coated onto small particles of gold or tungsten approximately two micrometres in diameter. The particles are placed in a vacuum chamber and the plant tissue to be engineered is placed below the chamber. The particles are propelled at high velocity using a short pulse of high pressure helium gas, and hit a fine mesh baffle placed above the tissue while the DNA coating continues into any target cell or tissue.

Agrobacterium method

Transformation via *Agrobacterium* has been successfully practiced in dicots, i.e. broadleaf plants, such as soybeans and tomatoes, for many years. Recently it has been adapted and is now effective in monocots like grasses, including corn and rice. In general, the *Agrobacterium* method is considered preferable to the gene gun, because of a greater frequency of single-site insertions of the foreign DNA, which allows for easier monitoring. In this method, the tumor inducing (Ti) region is removed from the T-DNA (transfer DNA) and replaced with the desired gene and a marker, which is then inserted into the organism. This may involve direct inoculation of the tissue with a culture of transformed Agrobacterium, or inoculation following treatment with micro-projectile bombardment, which wounds the tissue. Wounding of the target tissue causes the release of phenolic compounds by the plant, which induces invasion of the tissue by Agrobacterium. Because of this, microprojectile bombardment often increases the efficiency of infection with Agrobacterium. The marker is used to find the organism which has successfully taken up the desired gene. Tissues of the organism are then transferred to a medium containing an antibiotic or herbicide, depending on which marker was used. The *Agrobacterium* present is also killed by the antibiotic. Only tissues expressing the marker will survive and possess the gene of interest. Thus, subsequent steps in the process will only use these surviving plants. In order to obtain whole plants from these tissues, they are grown under controlled environmental conditions in tissue culture. This is a process of a series of media, each containing nutrients and hormones. Once the plants are grown and produce seed, the process of evaluating the progeny begins. This process entails selection of the seeds with the desired traits and then retesting and growing to make sure that the entire process has been completed successfully with the desired results.

GENETICALLY ENGINEERED CROPS

Genetically Engineered Crops

The use of genetically engineered crops has helped many farmers deal with pest problems that reduce their crop production. The impact of pest-resistant crops has led to a much higher yield for farmers in today's world. They can use less pesticides which reduces the chemicals that they put into the ground. Certain engineered crops have allowed farmers in the United States and worldwide increase crop yield exponentially in recent years. armers can use a glyphosate herbicide to kill weeds, yet the genetically engineered corn is resistant to the herbicide and is left unaffected. Thus, fields are produced that are virtually weed free. Genetically engineered crops can also benefit farmers when dealing with potentially harmful viruses and bacteria. In the 1990s a mutant strain of virus was decimating the commercial corn fields of the United States. Scientists found a virus-resistant strain of maize in the highlands of Mexico and extracted the part of the maize's genome that coded for resistance against the virus and incorporated it into their existing strain of commercial corn. This allowed the commercial strain to produce progeny that were resistant to the virus. Thus, the crops were saved from decimation.

Potential Detrimental Effects of Genetically Engineered Plants

According to John E. Berringer the outcome of releasing genetically modified organisms into the environment is still not known.

REFERENCES

1. Brooker, Robert. Genetics Analysis and Principles. 3rd. New York: McGraw-Hill Irwin, 2009. Print.
2. Stern, Kingsley. Introductory Plant Biology. 11th. New York City: McGraw-Hill, 2008.
3. "Expression of an Arabidopsis sodium/proton antiporter gene (AtNHX1) in peanut to improve salt tolerance - Springer". Plant Biotechnology Reports. Link. springer.com. 6: 59–67. 2012-01-01. doi:10.1007/s11816-011-0200-5. Retrieved 2014-02-06.
4. Bidney, D; Scelonge, C; Martich, J; Burrus, M; Sims, L; Huffman, G (January 1992). "Microprojectile bombardment of plant tissues increases transformation frequency by Agrobacterium tumefaciens". Plant Mol. Biol. 18 (2): 301–13. doi:10.1007/bf00034957. PMID 1310058.
5. Releasing Genetically Modified Organisms: Will Any Harm Outweigh Any Advantage? John E. Beringer. The Journal of Applied Ecology, Vol. 37, No. 2 (Apr., 2000), pp. 207-214. Published by: British Ecological Society

CHAPTER 2

Chromatin, Centromere, and Telomeres

Chapter Contents

Chromatin is a mass of genetic material composed of DNA and proteins that condenses to form chromosomes during eukaryotic cell division. Chromatin is located in the nucleus of cells. The primary function of chromatin is to compress the DNA into a compact unit that will be less voluminous and can fit within the nucleus. Chromatin consists of complexes of small proteins known as histones and DNA. Histones help to organize DNA into structures called nucleosomes by providing a base on which the DNA can be wrapped around. A nucleosome consists of a DNA sequence of about 150 base pairs that is wrapped around a set of eight histones called an octomer. The nucleosome is further folded to produce a chromatin fiber. Chromatin fibers are coiled and condensed to form chromosomes.

INTRODUCTION

Chromatin is a complex of macromolecules found in cells, consisting of DNA, protein, and RNA. The primary functions of chromatin are 1) to package DNA into a more compact, denser shape, 2) to reinforce the DNA macromolecule to allow mitosis, 3) to prevent DNA damage, and 4) to control gene expression and DNA replication. The primary protein components of chromatin are histones that compact the DNA. Chromatin is only found in eukaryotic cells (cells with defined nuclei). Prokaryotic cells have a different organization of their DNA (the prokaryotic chromosome equivalent is called genophore and is localized within the nucleoid region).

The structure of chromatin depends on several factors. The overall structure depends on the stage of the cell cycle. During interphase, the chromatin is structurally loose to allow access to RNA and DNA polymerases that transcribe and replicate the DNA. The local structure of chromatin during interphase depends on the genes present on the DNA. That DNA which codes genes that are actively transcribed ("turned on") is more loosely packaged and associated with RNA polymerases (referred to as euchromatin) while that DNA which codes inactive genes ("turned off") is more condensed and associated with structural proteins (heterochromatin). Epigenetic chemical modification of the structural proteins in chromatin also alters the local chromatin structure, in particular chemical modifications of histone proteins by methylation and acetylation. As the cell prepares to divide, i.e. enters mitosis or meiosis, the chromatin packages more tightly to facilitate segregation of the chromosomes during anaphase. During this stage of the cell cycle this makes the individual chromosomes in many cells visible by optical microscope.

In general terms, there are three levels of chromatin organization:

- DNA wraps around histone proteins forming nucleosomes; the "beads on a string" structure (euchromatin).
- Multiple histones wrap into a 30 nm fibre consisting of nucleosome arrays in their most compact form (heterochromatin). (Definitively established to exist in vitro, the 30-nanometer fibre was not seen in recent X-ray studies of human mitotic chromosomes.)
- Higher-level DNA packaging of the 30 nm fibre into the metaphase chromosome (during mitosis and meiosis).

There are, however, many cells that do not follow this organisation. For example, spermatozoa and avian red blood cells have more tightly packed chromatin than most eukaryotic cells, and trypanosomatid protozoa do not condense their chromatin into visible chromosomes for mitosis.

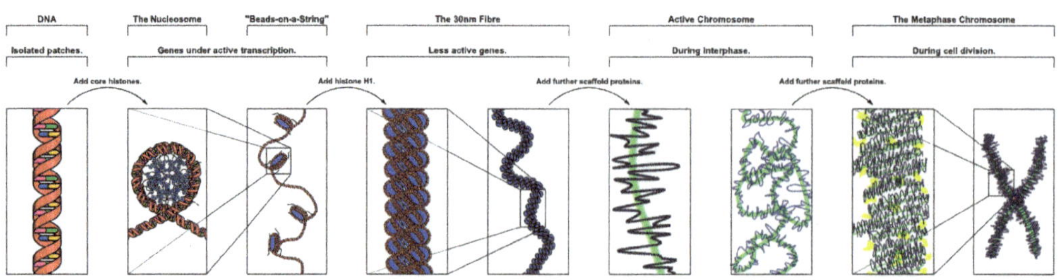

The major structures in DNA compaction: DNA, the nucleosome, the 10 nm "beads-on-a-string" fibre, the 30 nm chromatin fibre and the metaphase chromosome.

DYNAMIC CHROMATIN STRUCTURE AND HIERARCHY

Chromatin undergoes various structural changes during a cell cycle. Histone proteins are the basic packer and arranger of chromatin and can be modified by various post-translational modifications to alter chromatin packing (Histone modification). Most of the modifications occur on the histone tail. The consequences in terms of chromatin accessibility and compaction depend both on the amino-acid that is modified and the type of modification. For example, Histone acetylation results in loosening and increased accessibility of chromatin for replication and transcription. Lysine tri-methylation can either be correlated with transcriptional activity (tri-methylation of histone H3 Lysine 4) or transcriptional repression and chromatin compaction (tri-methylation of histone H3 Lysine 9 or 27). Several studies suggested that different modifications could occur simultaneously. For example, it was proposed that a bivalent structure (with tri-methylation of both Lysine 4 and 27 on histone H3) was involved in mammalian early development.

Polycomb-group proteins play a role in regulating genes through modulation of chromatin structure.

For additional information, see Histone modifications in chromatin regulation and RNA polymerase control by chromatin structure.

DNA Structure

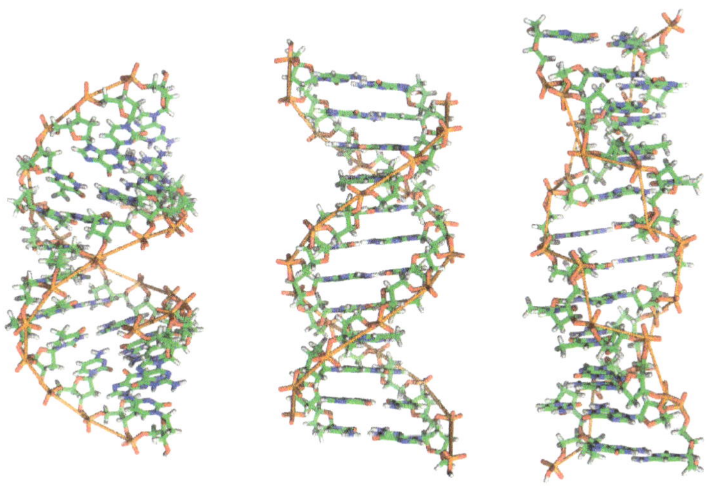

The structures of A-, B-, and Z-DNA.

In nature, DNA can form three structures, A-, B-, and Z-DNA. A- and B-DNA are very similar, forming right-handed helices, whereas Z-DNA is a left-handed helix with a zig-zag phosphate backbone. Z-DNA is thought to play a specific role in chromatin structure and transcription because of the properties of the junction between B- and Z-DNA.

At the junction of B- and Z-DNA, one pair of bases is flipped out from normal bonding. These play a dual role of a site of recognition by many proteins and as a sink for torsional stress from RNA polymerase or nucleosome binding.

Nucleosomes and Beads-on-a-String

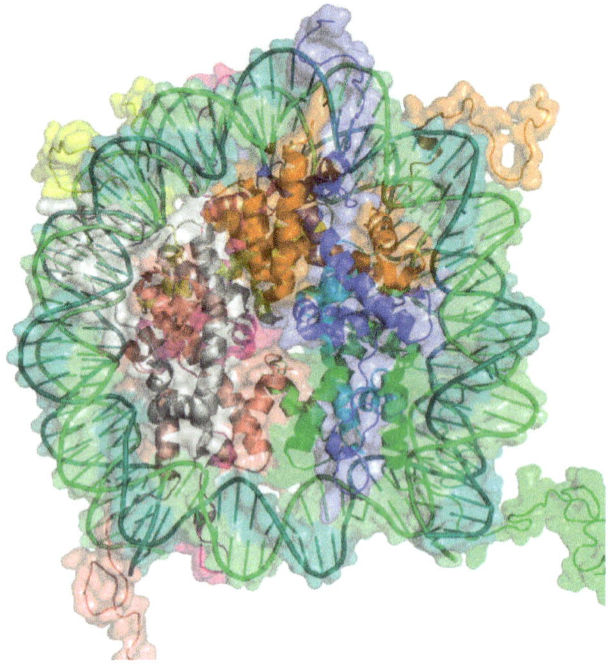

A cartoon representation of the nucleosome structure. From PDB: 1KX5.

The basic repeat element of chromatin is the nucleosome, interconnected by sections of linker DNA, a far shorter arrangement than pure DNA in solution.

In addition to the core histones, there is the linker histone, H1, which contacts the exit/entry of the DNA strand on the nucleosome. The nucleosome core particle, together with histone H1, is known as a chromatosome. Nucleosomes, with about 20 to 60 base pairs of linker DNA, can form, under non-physiological conditions, an approximately 10 nm "beads-on-a-string" fibre.

The nucleosomes bind DNA non-specifically, as required by their function in general DNA packaging. There are, however, large DNA sequence preferences that govern nucleosome positioning. This is due primarily to the varying physical properties of different DNA sequences: For instance, adenine and thymine are more favorably compressed into the inner minor grooves. This means nucleosomes can bind preferentially at one position approximately every 10 base pairs (the helical repeat of DNA)- where the DNA is rotated to maximise the number of A and T bases that will lie in the inner minor groove. (mechanical properties of DNA.)

30 Nanometer Chromatin Fibre

Two proposed structures of the 30nm chromatin filament. Left: 1 start helix "solenoid" structure. Right: 2 start loose helix structure. Note: the histones are omitted in this diagram - only the DNA is shown.

With addition of H1, the beads-on-a-string structure in turn coils into a 30 nm diameter helical structure known as the 30 nm fibre or filament. The precise structure of the chromatin fibre in the cell is not known in detail, and there is still some debate over this.

This level of chromatin structure is thought to be the form of heterochromatin, which contains mostly transcriptionally silent genes. EM studies have demonstrated that the 30 nm fibre is highly dynamic such that it unfolds into a 10 nm fiber ("beads-on-a-string") structure when transversed by an RNA polymerase engaged in transcription.

Four proposed structures of the 30 nm chromatin filament for DNA repeat length per nucleosomes ranging from 177 to 207 bp. Linker DNA in yellow and nucleosomal DNA in pink.

The existing models commonly accept that the nucleosomes lie perpendicular to the axis of the fibre, with linker histones arranged internally. A stable 30 nm fibre relies on the regular positioning of nucleosomes along DNA. Linker DNA is relatively resistant to bending and rotation. This makes the length of linker DNA critical to the stability of the fibre, requiring nucleosomes to be separated by lengths that permit rotation and folding into the required orientation without excessive stress to the DNA. In this view, different lengths of the linker DNA should produce different folding topologies of the chromatin fiber. Recent theoretical work, based on electron-microscopy images of reconstituted fibers supports this view.

Spatial Organization of Chromatin in the Cell Nucleus

The spatial arrangement of the chromatin within the nucleus is not random - specific regions of the chromatin can be found in certain territories. Territories are, for example, the lamina-associated domains (LADs), and the topological association domains (TADs), which are bound together by protein complexes. Currently, polymer models such as the Strings & Binders Switch (SBS) model and the Dynamic Loop (DL) model are used to describe the folding of chromatin within the nucleus.

Cell-Cycle Dependent Structural Organization

- **Interphase**: The structure of chromatin during interphase of mitosis is optimized to allow simple access of transcription and DNA repair factors to the DNA while compacting the DNA into the nucleus. The structure varies depending on the access required to the DNA. Genes that require regular access by RNA polymerase require the looser structure provided by euchromatin.

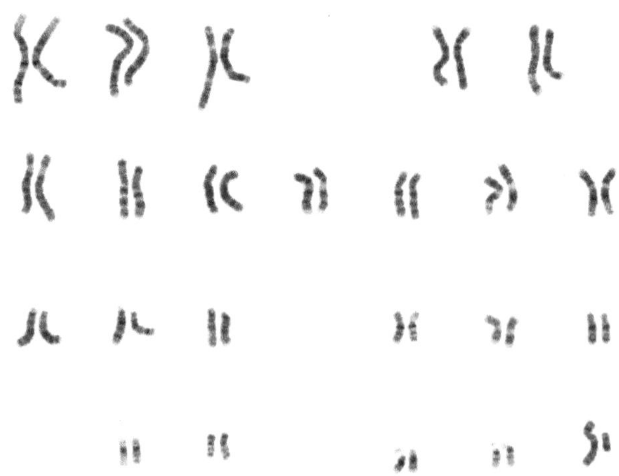

Karyogram of human male using Giemsa staining, showing the classic metaphase chromatin structure.

- **Metaphase**: The metaphase structure of chromatin differs vastly to that of interphase. It is optimised for physical strength and manageability, forming the classic chromosome structure seen in karyotypes. The structure of the condensed chromatin is thought to be loops of 30 nm fibre to a central scaffold of proteins. It is, however, not well-characterised.The physical strength of chromatin is vital for this stage of division to prevent shear damage to the DNA as the daughter chromosomes are separated. To maximise strength the composition of the chromatin changes as it approaches the centromere, primarily through alternative histone H1 analogues.It should also be noted that, during mitosis, while most of the chromatin is tightly compacted, there are small regions that are not as tightly compacted. These regions often correspond to promoter regions of genes that were active in that cell type prior to entry into chromatosis. The lack of compaction of these regions is called bookmarking, which is an epigenetic mechanism believed to be important for transmitting to daughter cells the "memory" of which genes were active prior to entry into mitosis. This bookmarking mechanism is needed to help transmit this memory because transcription ceases during mitosis.

Chromatin and Bursts of Transcription

Chromatin and its interaction with enzymes has been researched, and a conclusion being made is that it is relevant and an important factor in gene expression. Vincent G. Allfrey, a professor at Rockefeller University, stated that RNA synthesis is related to histone acetylation. The lysine amino acid attached to the end of the histones is positively charged. The acetylation of these tails would make the chromatin ends neutral, allowing for DNA access.

When the chromatin decondenses, the DNA is open to entry of molecular machinery. Fluctuations between open and closed chromatin may contribute to the discontinuity of transcription, or transcriptional bursting. Other factors are probably involved, such as the association and dissociation of transcription factor complexes with chromatin. The phenomenon, as opposed to simple probabilistic models of transcription, can account for the high variability in gene expression occurring between cells in isogenic populations

Alternative Chromatin Organizations

During metazoan spermiogenesis, the spermatid's chromatin is remodeled into a more spaced-packaged, widened, almost crystal-like structure. This process is associated with the cessation of transcription and involves nuclear protein exchange. The histones are mostly displaced, and replaced by protamines (small, arginine-rich proteins). It is proposed that in yeast, regions devoid of histones become very frigile after transcription; HMO1 an HMGB protein helps in stabilizing nucleosomes-free chromatin.

Methods to Investigate Chromatin

- **ChIP-seq** (Chromatin immunoprecipitation sequencing), aimed against different histone modifications, can be used to identify chromatin states throughout the genome. Different modifications have been linked to various states of chromatin.

- **DNase-seq** (DNase I hypersensitive sites Sequencing) uses the sensitivity of accessible regions in the genome to the DNase I enzyme to map open or accessible regions in the genome.
- **FAIRE-seq** (Formaldehyde-Assisted Isolation of Regulatory Elements sequencing) uses the chemical properties of protein-bound DNA in a two-phase separation method to extract nucleosome depleted regions from the genome.
- **ATAC-seq** (Assay for Transposable Accessible Chromatin sequencing) uses the Tn5 transposase to integrate (synthetic) transposons into accessible regions of the genome consequentially highlighting the localisation of nucleosomes and transcription factors across the genome.
- **DNA footprinting** is a method aimed at identifying protein-bound DNA. It uses labeling and fragmentation coupled to gel electrophoresis to identify areas of the genome that have been bound by proteins.
- **MNase-seq** (Micrococcal Nuclease sequencing) uses the micrococcal nuclease enzyme to identify nucleosome positioning throughout the genome.
- **Chromosome conformation capture** determines the spatial organization of chromatin in the nucleus, by inferring genomic locations that physically interact.
- **MACC profiling** (Micrococcal nuclease ACCessibility profiling) uses titration series of chromatin digests with micrococcal nuclease to identify chromatin accessibility as well as to map nucleosomes and non-histone DNA-binding proteins in both open and closed regions of the genome.

Chromatin: Alternative Definitions

The term, introduced by Walther Flemming, has multiple meanings:

- **Simple and concise definition:** Chromatin is a macromolecular complex of a DNA macromolecule and protein macromolecules (and RNA). The proteins package and arrange the DNA and control its functions within the cell nucleus.
- **A biochemists' operational definition:** Chromatin is the DNA/protein/RNA complex extracted from eukaryotic lysed interphase nuclei. Just which of the multitudinous substances present in a nucleus will constitute a part of the extracted material partly depends on the technique each researcher uses. Furthermore, the composition and properties of chromatin vary from one cell type to the another, during development of a specific cell type, and at different stages in the cell cycle.
- **The *DNA + histone = chromatin* definition:** The DNA double helix in the cell nucleus is packaged by special proteins termed histones. The formed protein/DNA complex is called chromatin. The basic structural unit of chromatin is the nucleosome.

Nobel Prizes

The following scientists were recognized for their contributions to chromatin research with Nobel Prizes:

Year	Who	Award
1910	Albrecht Kossel (University of Heidelberg)	Nobel Prize in Physiology or Medicine for his discovery of the five nuclear bases: adenine, cytosine, guanine, thymine, and uracil.
1933	Thomas Hunt Morgan (California Institute of Technology)	Nobel Prize in Physiology or Medicine for his discoveries of the role played by the gene and chromosome in heredity, based on his studies of the white-eyed mutation in the fruit fly *Drosophila*.
1962	Francis Crick, James Watson and Maurice Wilkins (MRC Laboratory of Molecular Biology, Harvard University and London University respectively)	Nobel Prize in Physiology or Medicine for their discoveries of the double helix structure of DNA and its significance for information transfer in living material.
1982	Aaron Klug (MRC Laboratory of Molecular Biology)	Nobel Prize in Chemistry "for his development of crystallographic electron microscopy and his structural elucidation of biologically important nucleic acid-protein complexes"
1993	Richard J. Roberts and Phillip A. Sharp	Nobel Prize in Physiology "for their independent discoveries of split genes," in which DNA sections called exons express proteins, and are interrupted by DNA sections called introns, which do not express proteins.
2006	Roger Kornberg (Stanford University)	Nobel Prize in Chemistry for his discovery of the mechanism by which DNA is transcribed into messenger RNA.

CENTROMERE

The **centromere** is the part of a chromosome that links sister chromatids or a dyad. During mitosis, spindle fibers attach to the centromere via the kinetochore. Centromeres were first thought to be genetic loci that direct the behavior of chromosomes.

The physical role of the centromere is to act as the site of assembly of the kinetochore - a highly complex multiprotein structure that is responsible for the actual events of chromosome segregation - i.e. binding microtubules and signalling to the cell cycle machinery when all chromosomes have adopted correct attachments to the spindle, so that it is safe for cell division to proceed to completion and for cells to enter anaphase.

There are, broadly speaking, two types of centromeres. "Point centromeres" bind to specific proteins that recognise particular DNA sequences with high efficiency. Any piece of DNA with the point centromere DNA sequence on it will typically form a centromere if present in the appropriate species. The best characterised point centromeres are those of the budding yeast, *Saccharomyces cerevisiae*. "Regional centromeres" is the term coined to describe most centromeres, which typically form on regions of preferred DNA sequence, but which can form on other DNA (Deoxyribonucleic acid) sequences as well. The signal for formation of a regional centromere appears to be epigenetic. Most organisms, ranging from the fission yeast *Schizosaccharomyces pombe* to humans, have regional centromeres.

Regarding mitotic chromosome structure, centromeres represent a constricted region of the chromosome (often referred to as the primary constriction) where two identical sister chromatids are most closely in contact. When cells enter mitosis, the sister chromatids (the two copies of each chromosomal DNA molecule resulting from DNA replication in chromatin form) are linked along their length by the action of the cohesin complex. It is now believed that this complex is mostly released from chromosome arms during prophase, so that by the time the chromosomes line up at the mid-plane of the mitotic spindle (also known as the metaphase plate), the last place where they are linked with one another is in the chromatin in and around the centromere.

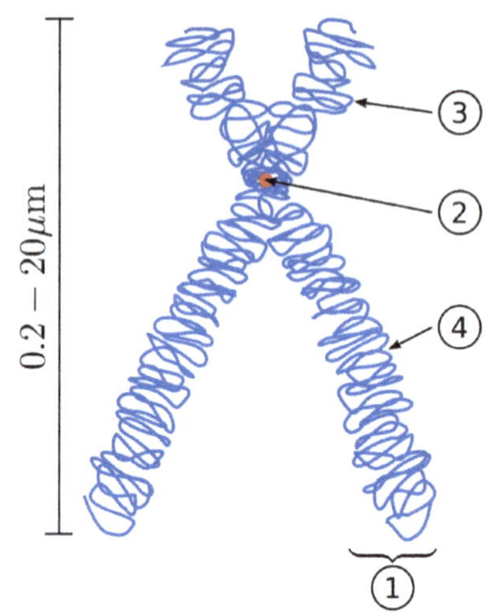

In this diagram of a duplicated chromosome, (2) identifies the centromere—the region that joins the two sister chromatids, or each half of the chromosome. In prophase of mitosis, specialized regions on centromeres called kinetochores attach chromosomes to spindle fibers.

Positions

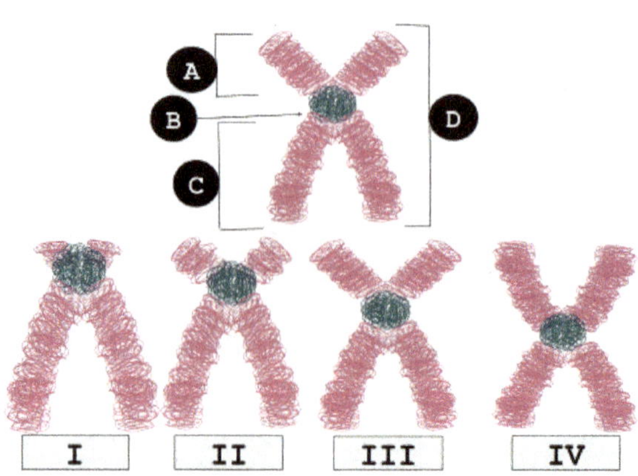

Classifications of Chromosomes

I: Telocentric - centromere placement very close to the top, p arms barely visible if visible at all

II: Acrocentric - q arms are still much longer than the p arms, but the p arms are longer than it those in telocentric

III: Submetacentric - p and q arms are very close in length but not equal

IV: Metacentric - the p arm and the q arms are equal in length

A: Short arm (p arm)

B: Centromere

C: Long arm (q arm)

D: Sister Chromatid

Each chromosome has two arms, labeled p (the shorter of the two) and q (the longer). Many remember that the short arm 'p' is named for the French word "petit" meaning 'small', although this explanation was shown to be apocryphal. They can be connected in either metacentric, submetacentric, acrocentric or telocentric manner.

Metacentric

These are X-shaped chromosomes, with the centromere in the middle so that the two arms of the chromosomes are almost equal.

A chromosome is metacentric if its two arms are roughly equal in length. In a normal human karyotype, five chromosomes are considered metacentric: chromosomes 1, 3, 16, 19, and 20. In some cases, a metacentric chromosome is formed by balanced translocation: the fusion of two acrocentric chromosomes to form one metacentric chromosome.

Submetacentric

If arms' lengths are unequal, the chromosome is said to be submetacentric.

Acrocentric

If the p (short) arm is so short that it is hard to observe, but still present, then the chromosome is acrocentric (the "acro-" in acrocentric refers to the Greek word for "peak"). The human genome includes six acrocentric chromosomes: 13, 14, 15, 21, 22 and the Y chromosome.

In an acrocentric chromosome the p arm contains genetic material including repeated sequences such as nucleolar organizing regions, and can be translocated without significant harm, as in a balanced Robertsonian translocation. The domestic horse genome includes one metacentric chromosome that is homologous to two acrocentric chromosomes in the conspecific but undomesticated Przewalski's horse. This may reflect either fixation of a balanced Robertsonian translocation in domestic horses or, conversely, fixation of the fission of one metacentric chromosome into two acrocentric chromosomes in Przewalski's horses. A similar situation exists between the human and great ape genomes; in this case, because more species are extant, it is apparent that the evolutionary sequence is a reduction of two acrocentric chromosomes in the great apes to one metacentric chromosome in humans (see Karyotype#Aneuploidy).

Telocentric

A telocentric chromosome's centromere is located at the terminal end of the chromosome. Telomeres may extend from both ends of the chromosome. For example, the standard house mouse karyotype has only acrocentric chromosomes. Humans do not possess telocentric chromosomes.

Subtelocentric

If the chromosome's centromere is located closer to its end than to its center, it may be described as subtelocentric.

Holocentric

With **holocentric** chromosomes, the entire length of the chromosome acts as the centromere. Examples of this type of centromere can be found scattered throughout the plant and animal kingdoms, with the most well-known example being the nematode *Caenorhabditis elegans*.

Human Chromosomes

Chromosome	Centromere position (Mbp)		Chromosome Size (Mbp)	Centromere size (Mbp)
1	125.0	metacentric	247.2	7.4
2	93.3	submetacentric	242.8	6.3
3	91.0	metacentric	199.4	6.0
4	50.4	submetacentric	191.3	
5	48.4	submetacentric	180.8	
6	61.0	submetacentric	170.9	
7	59.9	submetacentric	158.8	
8	45.6	submetacentric	146.3	
9	49.0	submetacentric	140.4	
10	40.2	submetacentric	135.4	
11	53.7	submetacentric	134.5	
12	35.8	submetacentric	132.3	
13	17.9	acrocentric	114.1	
14	17.6	acrocentric	106.3	
15	19.0	acrocentric	100.3	

Chromosome	Centromere position (Mbp)		Chromosome Size (Mbp)	Centromere size (Mbp)
16	36.6	metacentric	88.8	
17	24.0	submetacentric	78.7	
18	17.2	submetacentric	76.1	
19	26.5	metacentric	63.8	
20	27.5	metacentric	62.4	
21	13.2	acrocentric	46.9	
22	14.7	acrocentric	49.5	
X	60.6	submetacentric	154.9	
Y	12.5	acrocentric	57.7	

Sequence

There are two types of centromeres. In regional centromeres, DNA sequences contribute to but do not define function. Regional centromeres contain large amounts of DNA and are often packaged into heterochromatin. In most eukaryotes, the centromere's DNA sequence consists of large arrays of repetitive DNA (e.g. satellite DNA) where the sequence within individual repeat elements is similar but not identical. In humans, the primary centromeric repeat unit is called α-satellite (or alphoid), although a number of other sequence types are found in this region.

Point centromeres are smaller and more compact. DNA sequences are both necessary and sufficient to specify centromere identity and function in organisms with point centromeres. In budding yeasts, the centromere region is relatively small (about 125 bp DNA) and contains two highly conserved DNA sequences that serve as binding sites for essential kinetochore proteins.

Inheritance

Since centromeric DNA sequence is not the key determinant of centromeric identity in metazoans, it is thought that epigenetic inheritance plays a major role in specifying the centromere. The daughter chromosomes will assemble centromeres in the same place as the parent chromosome, independent of sequence. It has been proposed that histone H3 variant CENP-A (Centromere Protein A) is the epigenetic mark of the centromere. The question arises whether there must be still some original way in which the centromere is specified, even if it is subsequently propagated epigenetically. If the centromere is inherited epigenetically from one generation to the next, the problem is pushed back to the origin of the first metazoans.

Structure

The centromeric DNA is normally in a heterochromatin state, which is essential for the recruitment of the cohesin complex that mediates sister chromatid cohesion after DNA replication as well as coordinating sister chromatid separation during anaphase. In this chromatin, the normal histone H3 is replaced with a centromere-specific variant, CENP-A in humans. The presence of CENP-A is believed to be important for the assembly of the kinetochore on the centromere. CENP-C has been shown to localise almost exclusively to these regions of CENP-A associated chromatin. In human cells, the histones are found to be most enriched for H4K20me3 and H3K9me3 which are known heterochromatic modifications.

In the yeast *Schizosaccharomyces pombe* (and probably in other eukaryotes), the formation of centromeric heterochromatin is connected to RNAi. In nematodes such as *Caenorhabditis elegans*, some plants, and the insect orders Lepidoptera and Hemiptera, chromosomes are "holocentric", indicating that there is not a primary site of microtubule attachments or a primary constriction, and a "diffuse" kinetochore assembles along the entire length of the chromosome.

Centromeric Aberrations

In rare cases in humans, neocentromeres can form at new sites on the chromosome. There are currently over 90 known human neocentromeres identified on 20 different chromosomes. The formation of a neocentromere must be coupled with the inactivation of the previous centromere, since chromosomes with two functional centromeres (Dicentric chromosome) will result in chromosome breakage during mitosis. In some unusual cases human neocentromeres have been observed to form spontaneously on fragmented chromosomes. Some of these new positions were originally euchromatic and lack alpha satellite DNA altogether.

Centromere proteins are also the autoantigenic target for some anti-nuclear antibodies, such as anti-centromere antibodies.

Centromere Dysfunction and Disease

It has been known that centromere misregulation contributes to mis-segregation of chromosomes, which is strongly related to cancer and abortion. Notably, overexpression of many centromere genes have been linked to cancer malignant phenotypes. Overexpression of these centromere genes is thought to increase genomic instability in cancers. Elevated genomic instability on one hand relates to malignant phenotypes; on the other hand, it makes the tumor cells more vulnerable to specific adjuvant therapies such as certain chemotherapies and radiotherapy.

Etymology and Pronunciation

The word *centromere* uses combining forms of *centro- and -mere*, yielding "central part", describing the centromere's location at the center of the chromosome.

REFERENCES

1. "Chromatin Network Home Page.". Retrieved 2008-11-18.
2. Dame, R.T. (May 2005). "The role of nucleoid-associated proteins in the organization and compaction of bacterial chromatin". Molecular Microbiology. 56 (4): 858–870. doi:10.1111/j.1365-2958.2005.04598.x. PMID 15853876.
3. Hansen, Jeffrey (March 2012). "Human mitotic chromosome structure: what happened to the 30-nm fibre?". The EMBO Journal. 31 (7): 1621–1623. doi:10.1038/emboj.2012.66. PMC 3321215 . PMID 22415369.
4. Bernstein, B.E., T.S. Mikkelsen, X. Xie, M. Kamal, D.J. Huebert, J. Cuff, B. Fry, A. Meissner, M. Wernig, K. Plath, R. Jaenisch, A. Wagschal, R. Feil, S.L. Schreiber & E.S. Lander (April 2006). "A bivalent chromatin structure marks key developmental genes in embryonic stem cells". Cell. 125 (2): 315–26. doi:10.1016/j.cell.2006.02.041. ISSN 0092-8674. PMID 16630819.
5. Portoso M, Cavalli G (2008). "The Role of RNAi and Noncoding RNAs in Polycomb Mediated Control of Gene Expression and Genomic Programming". RNA and the Regulation of Gene Expression: A Hidden Layer of Complexity. Caister Academic Press. ISBN 978-1-904455-25-7.
6. Annunziato, Anthony T. "DNA Packaging: Nucleosomes and Chromatin". Scitable. Nature Education. Retrieved 2015-10-29.
7. Robinson DJ; Fairall L; Huynh VA; Rhodes D. (April 2006). "EM measurements define the dimensions of the "30-nm" chromatin fiber: Evidence for a compact, interdigitated structure". PNAS. 103 (17): 6506–11. doi:10.1073/pnas.0601212103. PMC 1436021 . PMID 16617109.
8. Wong H, Victor JM, Mozziconacci J. (September 2007). Chen, Pu, ed. "An All-Atom Model of the Chromatin Fiber Containing Linker Histones Reveals a Versatile Structure Tuned by the Nucleosomal Repeat Length". PLoS ONE. 2 (9): e877. doi:10.1371/journal.pone.0000877. PMC 1963316 . PMID 17849006.
9. Nicodemi M, Pombo A (June 2014). "Models of chromosome structure". Curr. Opin. Cell Biol. 28: 90–5. doi:10.1016/j.ceb.2014.04.004. PMID 24804566.
10. Nicodemi M, Panning B, Prisco A (May 2008). "A thermodynamic switch for chromosome colocalization". Genetics. 179 (1): 717–21. doi:10.1534/genetics.107.083154. PMC 2390650 . PMID 18493085.
11. Bohn M, Heermann DW (2010). "Diffusion-driven looping provides a consistent framework for chromatin organization". PLoS ONE. 5 (8): e12218. doi:10.1371/journal.pone.0012218. PMC 2928267 . PMID 20811620.
12. Xing H, Vanderford NL, Sarge KD (November 2008). "The TBP-PP2A mitotic complex bookmarks genes by preventing condensin action". Nat. Cell Biol. 10 (11): 1318–23. doi:10.1038/ncb1790. PMC 2577711 . PMID 18931662.
13. ALLFREY VG, FAULKNER R, MIRSKY AE (May 1964). "ACETYLATION AND METHYLATION OF HISTONES AND THEIR POSSIBLE ROLE IN THE REGULATION OF RNA SYNTHESIS". Proc. Natl. Acad. Sci. U.S.A. 51 (5): 786–94. doi:10.1073/pnas.51.5.786. PMC 300163 . PMID 14172992.

14. Kaochar S, Tu BP (November 2012). "Gatekeepers of chromatin: Small metabolites elicit big changes in gene expression". Trends Biochem. Sci. 37 (11): 477–83. doi:10.1016/j.tibs.2012.07.008. PMC 3482309 . PMID 22944281.

15. De Vries M, Ramos L, Housein Z, De Boer P (May 2012). "Chromatin remodelling initiation during human spermiogenesis". Biol Open. 1 (5): 446–57. doi:10.1242/bio.2012844. PMC 3507207 . PMID 23213436.

16. D. Murugesapillai et al, DNA bridging and looping by HMO1 provides a mechanism for stabilizing nucleosome-free chromatin, Nucl Acids Res (2014) 42 (14): 8996-9004

17. D. Murugesapillai et al, Single-molecule studies of high-mobility group B architectural DNA bending proteins, Biophys Rev (2016) doi:10.1007/s12551-016-0236-4

18. Giresi, Paul G.; Kim, Jonghwan; McDaniell, Ryan M.; Iyer, Vishwanath R.; Lieb, Jason D. (2007-06-01). "FAIRE (Formaldehyde-Assisted Isolation of Regulatory Elements) isolates active regulatory elements from human chromatin". Genome Research. 17 (6): 877–885. doi:10.1101/gr.5533506. ISSN 1088-9051. PMC 1891346 . PMID 17179217.

19. Galas, D. J.; Schmitz, A. (1978-09-01). "DNAse footprinting: a simple method for the detection of protein-DNA binding specificity". Nucleic Acids Research. 5 (9): 3157–3170. doi:10.1093/nar/5.9.3157. ISSN 0305-1048. PMC 342238 . PMID 212715.

20. Cui, Kairong; Zhao, Keji (2012-01-01). "Genome-wide approaches to determining nucleosome occupancy in metazoans using MNase-Seq". Methods in Molecular Biology (Clifton, N.J.). 833: 413–419. doi:10.1007/978-1-61779-477-3_24. ISSN 1940-6029. PMC 3541821 . PMID 22183607.

21. Buenrostro, Jason D.; Giresi, Paul G.; Zaba, Lisa C.; Chang, Howard Y.; Greenleaf, William J. (2013-12-01). "Transposition of native chromatin for fast and sensitive epigenomic profiling of open chromatin, DNA-binding proteins and nucleosome position". Nature Methods. 10 (12): 1213–1218. doi:10.1038/nmeth.2688. ISSN 1548-7105. PMC 3959825 . PMID 24097267.

22. Mieczkowski J, Cook A, Bowman SK, Mueller B, Alver BH, Kundu S, Deaton AM, Urban JA, Larschan E, Park PJ, Kingston RE, Tolstorukov MY (2016-05-06). "MNase titration reveals differences between nucleosome occupancy and chromatin accessibility.". Nature Communications. 7: 11485. doi:10.1038/ncomms11485. PMC 4859066 . PMID 27151365.

23. "Thomas Hunt Morgan and His Legacy". Nobelprize.org. 7 Sep 2012

24. Pollard, T.D. (2007). Cell Biology. Philadelphia: Saunders. pp. 200–203. ISBN 978-1-4160-2255-8.

25. Pollard, TD (2007). Cell Biology. Philadelphia: Saunders. pp. 227–230. ISBN 978-1-4160-2255-8.

26. Pluta, A.; A.M. Mackay; A.M. Ainsztein; I.G. Goldberg; W.C. Earnshaw (1995). "The centromere: Hub of chromosomal activities.". Science. 270 (5242): 1591–1594. doi:10.1126/science.270.5242.1591. PMID 7502067.

27. "Sister chromatid cohesion". Genetics Home Reference. United States National Library of Medicine. May 15, 2011.
28. "p + q = Solved, Being the True Story of How the Chromosome Got Its Name".
29. "Chromosomes, Chromosome Anomalies".
30. *Gilbert F (1999). "Disease genes and chromosomes: disease maps of the human genome. Chromosome 16". Genet Test. 3 (2): 243–54. PMID 10464676.
31. http://www.amazon.com/Thompson-Genetics-Medicine-Sixth-Edition/dp/0721669026
32. Myka, J.L.; Lear, T.L.; Houck, M.L.; Ryder, O.A.; Bailey, E. (2003). "FISH analysis comparing genome organization in the domestic horse (Equus caballus) to that of the Mongolian wild horse (E. przewalskii)". Cytogenetic and Genome Research. 102 (1–4): 222–5. doi:10.1159/000075753. PMID 14970707.
33. Silver, Lee M. (1995). "Karyotypes, Chromosomes, and Translocations". Mouse Genetics: Concepts and Applications. Oxford: Oxford University Press. pp. 83–92. ISBN 978-0-19-507554-0.
34. Chinwalla, Asif T.; Cook, Lisa L.; Delehaunty, Kimberly D.; Fewell, Ginger A.; Fulton, Lucinda A.; Fulton, Robert S.; Graves, Tina A.; Hillier, Ladeana W.; et al. (2002). "Initial sequencing and comparative analysis of the mouse genome". Nature. 420 (6915): 520–62. doi:10.1038/nature01262. PMID 12466850.
35. Margulis, Lynn; Matthews, Clifford; Haselton, Aaron (2000-01-01). Environmental Evolution: Effects of the Origin and Evolution of Life on Planet Earth. MIT Press. ISBN 9780262631976.
36. Dernburg, A. F. (2001). "Here, There, and Everywhere: Kinetochore Function on Holocentric Chromosomes". The Journal of Cell Biology. 153 (6): F33–8. doi:10.1083/jcb.153.6.F33. PMC 2192025 . PMID 11402076.
37. Pluta, A. F.; MacKay, A. M.; Ainsztein, A. M.; Goldberg, I. G.; Earnshaw, W. C. (1995). "The Centromere: Hub of Chromosomal Activities". Science. 270 (5242): 1591–4. doi:10.1126/science.270.5242.1591. PMID 7502067.
38. Mehta, G. D.; Agarwal, M.; Ghosh, S. K. (2010). "Centromere Identity: a challenge to be faced". Mol. Genet. Genomics. 284 (2): 75–94. doi:10.1007/s00438-010-0553-4. PMID 20585957.
39. Dalal, Yamini (2009). "Epigenetic specification of centromeres". Biochemistry and Cell Biology. 87 (1): 273–82. doi:10.1139/O08-135. PMID 19234541.
40. Bernad, Rafael; Sánchez, Patricia; Losada, Ana (2009). "Epigenetic specification of centromeres by CENP-A". Experimental Cell Research. 315 (19): 3233–41. doi:10.1016/j.yexcr.2009.07.023. PMID 19660450.
41. Chueh, A. C.; Wong, LH; Wong, N; Choo, KH (2004). "Variable and hierarchical size distribution of L1-retroelement-enriched CENP-A clusters within a functional human neocentromere". Human Molecular Genetics. 14 (1): 85–93. doi:10.1093/hmg/ddi008. PMID 15537667.
42. Rosenfeld, Jeffrey A; Wang, Zhibin; Schones, Dustin E; Zhao, Keji; Desalle, Rob; Zhang, Michael Q (2009). "Determination of enriched histone modifications

in non-genic portions of the human genome". BMC Genomics. 10: 143. doi:10.1186/1471-2164-10-143. PMC 2667539 . PMID 19335899.

43. Volpe, T. A.; Kidner, C; Hall, IM; Teng, G; Grewal, SI; Martienssen, RA (2002). "Regulation of Heterochromatic Silencing and Histone H3 Lysine-9 Methylation by RNAi". Science. 297 (5588): 1833–7. doi:10.1126/science.1074973. PMID 12193640.

44. Marshall, Owen J.; Chueh, Anderly C.; Wong, Lee H.; Choo, K.H. Andy (2008). "Neocentromeres: New Insights into Centromere Structure, Disease Development, and Karyotype Evolution". The American Journal of Human Genetics. 82 (2): 261–82. doi:10.1016/j.ajhg.2007.11.009. PMC 2427194 . PMID 18252209.

45. Warburton, Peter E. (2004). "Chromosomal dynamics of human neocentromere formation". Chromosome Research. 12 (6): 617–26. doi:10.1023/B:CHRO.0000036585.44138.4b. PMID 15289667.

46. Zhang, W.; Mao, J-H.; Zhu, W.; Jain, A.K.; Liu, L.; Brown, J.B.; Karpen, G.H. (2016). "Centromere and kinetochore gene misexpression predicts cancer patient survival and response to radiotherapy and chemotherapy". Nature Communications. 7: 12619. doi:10.1038/ncomms12619. PMC 5013662 . PMID 27577169.

47. "Centromere". Merriam-Webster Dictionary.

48. "Centromere". Dictionary.com Unabridged. Random House.

CHAPTER 3

Genomics and Plant Genomes

Chapter Contents

Genomics is the study of the entire set of genes found in living things, including humans, plants, animals and even viruses. More specifically, it refers to the analysis of the genomes of organisms, both anatomically (sequences and organization) and physiologically (expression and regulation). The genome, a combination of the words "gene" and "chromosome," is the complete set of hereditary instructions in each cell of every living thing that are needed by the organism for its development and functioning. It is similar to an instruction manual on how cells work.

INTRODUCTION

Genomics is an interdisciplinary field of science focusing on genomes. A genome is a complete set of DNA within a single cell of an organism, and as such genomics is a branch of molecular biology concerned with the structure, function, evolution, and mapping of genomes. Genomics aims at the collective characterization and quantification of genes, which direct the production of proteins with the assistance of enzymes and messenger molecules. Proteins in turn make up body structures like organs and tissues as well as control chemical reactions and carry signals between cells. If a cell's DNA is mutated, an abnormal protein may be produced, which can disrupt the body's usual processes and lead to a disease such as cancer. In contrast to genetics, which refers to the study of genes and their roles in inheritance, genomics is the study of genes, their functions, and related techniques, such as applications of recombinant DNA, DNA sequencing methods, and bioinformatics to sequence, assemble, and analyze the function and structure of genomes. Advances in genomics have triggered a revolution in discovery-based research to understand even the most complex biological systems such as the brain. The field includes efforts to determine the entire DNA sequence of organisms and fine-scale genetic mapping. The field also includes studies of intragenomic phenomena such as heterosis, epistasis, pleiotropy and other interactions between loci and alleles within the genome. Research carried out into single genes does not generally fall into the definition of genomics unless the aim of this genetic, pathway, and functional information analysis is to elucidate its effect on, place in, and response to the entire genomes networks.

Etymology

From the Greek ΓEN *gen*, "gene" (gamma, epsilon, nu, epsilon) meaning "become, create, creation, birth", and subsequent variants: genealogy, genesis, genetics, genic, genomere, genotype, genus etc. While the word *genome* (from the German *Genom*, attributed to Hans Winkler) was in use in English as early as 1926, the term *genomics* was coined by Tom Roderick, a geneticist at the Jackson Laboratory (Bar Harbor, Maine), over beer at a meeting held in Maryland on the mapping of the human genome in 1986.

Early Sequencing Efforts

Following Rosalind Franklin's confirmation of the helical structure of DNA, James D. Watson and Francis Crick's publication of the structure of DNA in 1953 and Fred Sanger's publication of the Amino acid sequence of insulin in 1955, nucleic acid sequencing became a major target of early molecular biologists. In 1964, Robert W. Holley and colleagues published the first nucleic acid sequence ever determined, the ribonucleotide sequence of alanine transfer RNA. Extending this work, Marshall Nirenberg and Philip Leder revealed the triplet nature of the genetic code and were able to determine the sequences of 54 out of 64 codons in their experiments. In 1972, Walter Fiers and his team at the Laboratory of Molecular Biology of the University of Ghent (Ghent, Belgium) were the first to determine the sequence of a gene: the gene for Bacteriophage MS2 coat protein. Fiers' group expanded on their MS2 coat protein work, determining the complete nucleotide-sequence of bacteriophage MS2-RNA (whose genome encodes just four genes in 3569 base pairs and Simian virus 40 in 1976 and 1978, respectively.

DNA-Sequencing Technology Developed

Frederick Sanger

Walter Gilbert

Frederick Sanger and Walter Gilbert shared half of the 1980 Nobel Prize in chemistry for independently developing methods for the sequencing of DNA.

In addition to his seminal work on the amino acid sequence of insulin, Frederick Sanger and his colleagues played a key role in the development of DNA sequencing techniques that enabled the establishment of comprehensive genome sequencing projects. In 1975, he and Alan Coulson published a sequencing procedure using DNA polymerase with radiolabelled nucleotides that he called the *Plus and Minus technique*. This involved two closely related methods that generated short oligonucleotides with defined 3' termini. These could be fractionated by electrophoresis on a polyacrylamide gel and visualised using autoradiography. The procedure could sequence up to 80 nucleotides in one go and was a big improvement, but was still very laborious. Nevertheless, in 1977 his group was able to sequence most of the 5,386 nucleotides of the single-stranded bacteriophageφX174, completing the first fully sequenced DNA-based genome. The refinement of the *Plus and Minus* method resulted in the chain-termination, or Sanger method (see below), which formed the basis of the techniques of DNA sequencing, genome mapping, data storage, and bioinformatic analysis most widely used in the following quarter-century of research. In the same year Walter Gilbert and Allan Maxam of Harvard University independently developed the Maxam-Gilbert method (also known as the *chemical method*) of DNA sequencing, involving the preferential cleavage of DNA at known bases, a less efficient method. For their groundbreaking work in the sequencing of nucleic acids, Gilbert and Sanger shared half the 1980 Nobel Prize in chemistry with Paul Berg (recombinant DNA).

Complete Genomes

The advent of these technologies resulted in a rapid intensification in the scope and speed of completion of genome sequencing projects. The first complete genome sequence of an eukaryotic organelle, the human mitochondrion (16,568 bp, about 16.6 kb kilobase), was reported in 1981,and the first chloroplast genomes followed in 1986. In 1992, the first eukaryotic chromosome, chromosome III of brewer's yeast *Saccharomyces cerevisiae* (315 kb) was sequenced. The first free-living organism to be sequenced was that of *Haemophilus influenzae* (1.8 Mb megabase) in 1995. The following year a consortium of researchers from laboratories across North America, Europe, and Japan announced the completion of the first complete genome sequence of a eukaryote, *S. cerevisiae* (12.1 Mb), and since then genomes have continued being sequenced at an exponentially growing pace. As of October 2011, the complete sequences are available for: 2,719 viruses, 1,115 archaea and bacteria, and 36 eukaryotes, of which about half are fungi.

Most of the microorganisms whose genomes have been completely sequenced are problematic pathogens, such as *Haemophilus influenzae*, which has resulted in a pronounced bias in their phylogenetic distribution compared to the breadth of microbial diversity. Of the other sequenced species, most were chosen because they were well-studied model organisms or promised to become good models. Yeast (*Saccharomyces cerevisiae*) has long been an important model organism for the eukaryotic cell, while the fruit fly *Drosophila melanogaster* has been a very important tool (notably in early pre-molecular genetics). The worm *Caenorhabditis elegans* is an often used simple model for multicellular organisms. The zebrafish *Brachydanio rerio* is used for many developmental studies on the molecular level, and the flower *Arabidopsis thaliana* is a model organism for flowering plants. The Japanese pufferfish (*Takifugu rubripes*) and the spotted green pufferfish (*Tetraodon nigroviridis*) are interesting because of their small and compact genomes, which contain very little noncoding DNA compared to most species. The mammals dog (*Canis familiaris*), brown rat (*Rattus norvegicus*), mouse (*Mus musculus*), and chimpanzee (*Pan troglodytes*) are all important model animals in medical research.

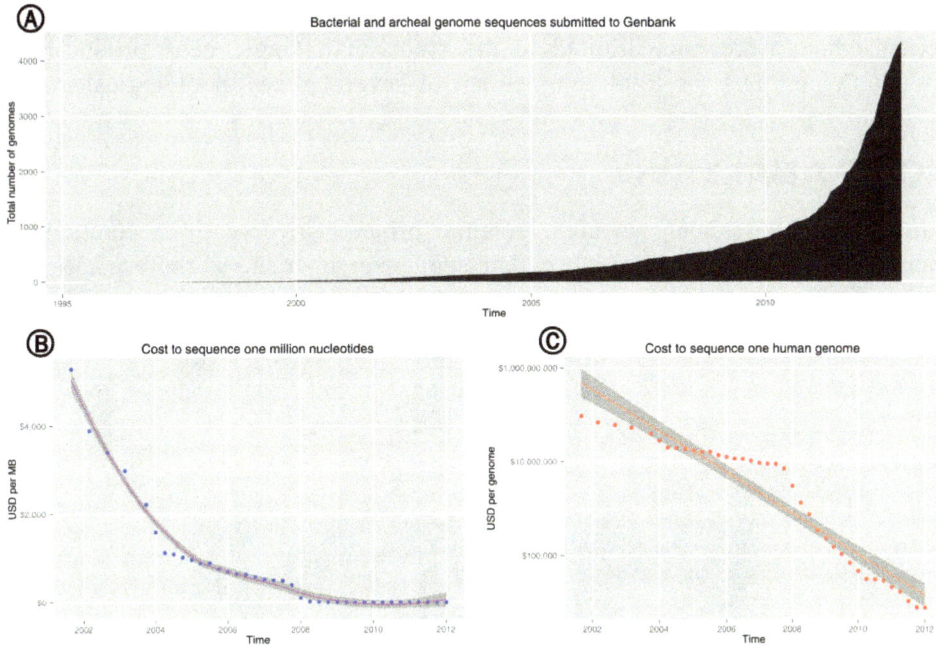

The number of genome projects has increased as technological improvements continue to lower the cost of sequencing. **(A)** Exponential growth of genome sequence databases since 1995. **(B)** The cost in US Dollars (USD) to sequence one million bases. **(C)** The cost in USD to sequence a 3,000 Mb (human-sized) genome on a log-transformed scale.

A rough draft of the human genome was completed by the Human Genome Project in early 2001, creating much fanfare. This project, completed in 2003, sequenced the entire genome for one specific person, and by 2007 this sequence was declared "finished" (less than one error in 20,000 bases and all chromosomes assembled). In the years since then, the genomes of many other individuals have been sequenced, partly under the auspices of the 1000 Genomes Project, which announced the sequencing of 1,092 genomes in October 2012.Completion of this project was made possible by the development of dramatically more efficient sequencing technologies and required the commitment of significant bioinformatics resources from a large international collaboration. The continued analysis of human genomic data has profound political and social repercussions for human societies.

The "Omics" Revolution

The English-language neologism **omics** informally refers to a field of study in biology ending in *-omics*, such as genomics, proteomics or metabolomics. The related suffix **-ome** is used to address the objects of study of such fields, such as the genome, proteome or metabolome respectively. The suffix *-ome* as used in molecular biology refers to a *totality* of some sort; similarly **omics** has come to refer generally to the study of large, comprehensive biological data sets. While the growth in the use of the term has led some scientists (Jonathan Eisen, among others) to claim that it has been oversold, it reflects the change in orientation towards the quantitative analysis of complete or near-complete assortment of all the constituents of a system. In the study of symbioses, for example, researchers which were once limited to the study of a single gene product can now simultaneously compare the total complement of several types of biological molecules.

GENOME ANALYSIS

After an organism has been selected, genome projects involve three components: the sequencing of DNA, the assembly of that sequence to create a representation of the original chromosome, and the annotation and analysis of that representation.

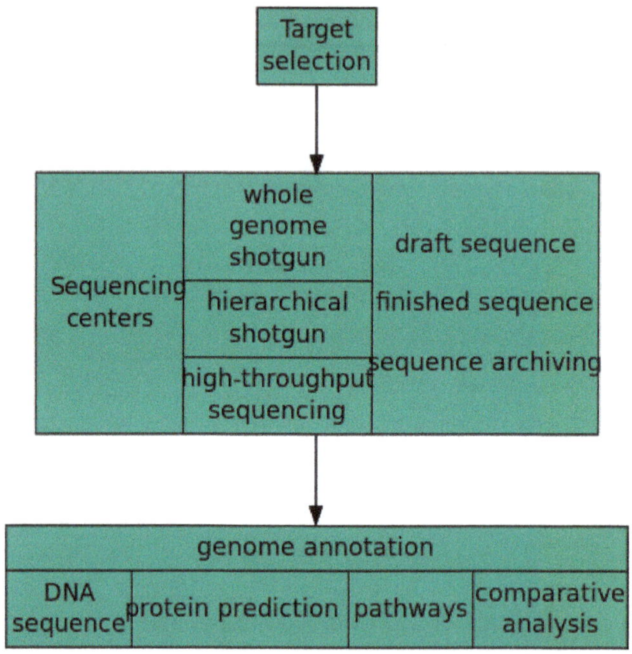

Overview of a genome project. First, the genome must be selected, which involves several factors including cost and relevance. Second, the sequence is generated and assembled at a given sequencing center (such as BGI or DOE JGI). Third, the genome sequence is annotated at several levels: DNA, protein, gene pathways, or comparatively.

Sequencing

Historically, sequencing was done in *sequencing centers*, centralized facilities (ranging from large independent institutions such as Joint Genome Institute which sequence dozens of terabases a year, to local molecular biology core facilities) which contain research laboratories with the costly instrumentation and technical support necessary. As sequencing technology continues to improve, however, a new generation of effective fast turnaround benchtop sequencers has come within reach of the average academic laboratory. On the whole, genome sequencing approaches fall into two broad categories, *shotgun* and *high-throughput* (aka *next-generation*) sequencing.

Shotgun Sequencing

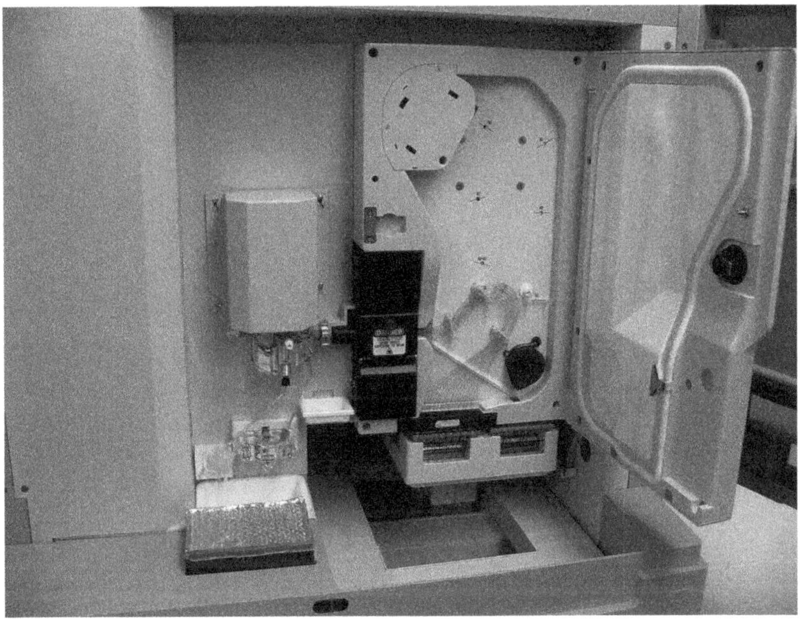

An ABI PRISM 3100 Genetic Analyzer. Such capillary sequencers automated early large-scale genome sequencing efforts.

Shotgun sequencing (Sanger sequencing is used interchangeably) is a sequencing method designed for analysis of DNA sequences longer than 1000 base pairs, up to and including entire chromosomes. It is named by analogy with the rapidly expanding, quasi-random firing pattern of a shotgun. Since the chain termination method of DNA sequencing can only be used for fairly short strands (100 to 1000 base pairs), longer DNA sequences must be broken into random small segments which are then sequenced to obtain *reads*. Multiple overlapping reads for the target DNA are obtained by performing several rounds of this fragmentation and sequencing. Computer programs then use the overlapping ends of different reads to assemble them into a continuous sequence. Shotgun sequencing is a random sampling process, requiring over-sampling to ensure a given nucleotide is represented in the reconstructed sequence; the average number of reads by which a genome is over-sampled is referred to as coverage.

For much of its history, the technology underlying shotgun sequencing was the classical chain-termination method, which is based on the selective incorporation of chain-terminating dideoxynucleotides by DNA polymerase during in vitro DNA replication. Developed by Frederick Sanger and colleagues in 1977, it was the most widely used sequencing method for approximately 25 years. More recently, Sanger sequencing has been supplanted by "Next-Gen" sequencing methods, especially for large-scale, automated genome analyses. However, the Sanger method remains in wide use in 2013, primarily for smaller-scale projects and for obtaining especially long contiguous DNA sequence reads (>500 nucleotides).Chain-termination methods require a single-stranded DNA template, a DNA primer, a DNA polymerase, normal deoxynucleosidetriphosphates (dNTPs), and modified nucleotides (dideoxyNTPs) that terminate DNA strand elongation. These chain-terminating nucleotides lack a 3'-OH group required for the formation of a phosphodiester bond between two nucleotides, causing DNA polymerase to cease extension of DNA when a ddNTP is incorporated. The ddNTPs may be radioactively or fluorescently labelled for detection in automated sequencing machines. Typically, these automated DNA-sequencing instruments (DNA sequencers) can sequence up to 96 DNA samples in a single batch (run) in up to 48 runs a day.

High-Throughput Sequencing

The high demand for low-cost sequencing has driven the development of high-throughput sequencing (or next-generation sequencing NGS) technologies that parallelize the sequencing process, producing thousands or millions of sequences at once. High-throughput sequencing technologies are intended to lower the cost of DNA sequencing beyond what is possible with standard dye-terminator methods. In ultra-high-throughput sequencing as many as 500,000 sequencing-by-synthesis operations may be run in parallel.

Illumina (Solexa) Sequencing

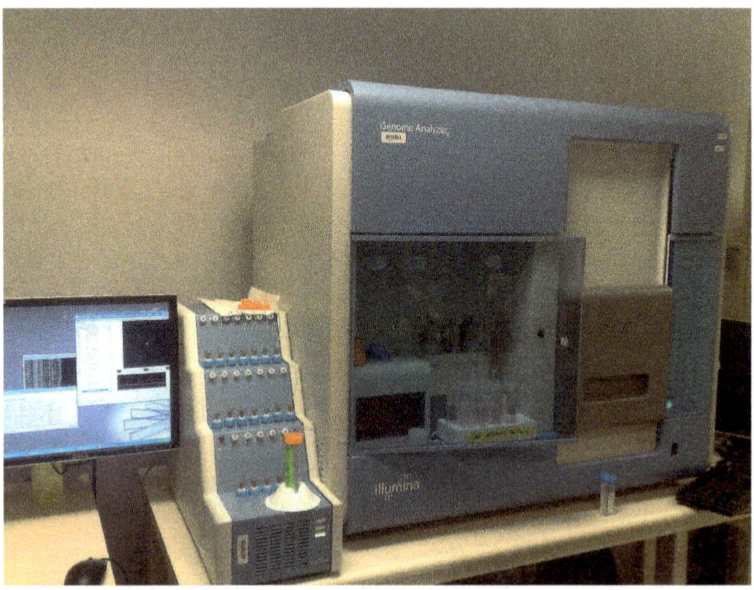

Illumina Genome Analyzer II System. Illumina technologies have set the standard for high throughput massively parallel sequencing.

Solexa, now part of Illumina, developed a sequencing method based on reversible dye-terminators technology acquired from Manteia Predictive Medicine in 2004. This technology had been invented and developed in late 1996 at Glaxo-Welcome's Geneva Biomedical Research Institute (GBRI), by Dr. Pascal Mayer and Dr Laurent Farinelli. In this method, DNA molecules and primers are first attached on a slide and amplified with polymerase so that local clonal colonies, initially coined "DNA colonies", are formed. To determine the sequence, four types of reversible terminator bases (RT-bases) are added and non-incorporated nucleotides are washed away. Unlike pyrosequencing, the DNA chains are extended one nucleotide at a time and image acquisition can be performed at a delayed moment, allowing for very large arrays of DNA colonies to be captured by sequential images taken from a single camera.

Decoupling the enzymatic reaction and the image capture allows for optimal throughput and theoretically unlimited sequencing capacity. With an optimal configuration, the ultimately reachable instrument throughput is thus dictated solely by the analogic-to-digital conversion rate of the camera, multiplied by the number of cameras and divided by the number of pixels per DNA colony required for visualizing them optimally (approximately 10 pixels/colony). In 2012, with cameras operating at more than 10 MHz A/D conversion rates and available optics, fluidics and enzymatics, throughput can be multiples of 1 million nucleotides/second, corresponding roughly to 1 human genome equivalent at 1x coverage per hour per instrument, and 1 human genome re-sequenced (at approx. 30x) per day per instrument (equipped with a single camera). The camera takes images of the fluorescently labeled nucleotides, then the dye along with the terminal 3' blocker is chemically removed from the DNA, allowing the next cycle.

Ion Torrent

Ion Torrent Systems Inc. developed a sequencing approach based on standard DNA replication chemistry. This technology measures the release of a hydrogen ion each time a base is incorporated. A microwell containing template DNA is flooded with a single nucleotide, if the nucleotide is complementary to the template strand it will be incorporated and a hydrogen ion will be released. This release triggers an ISFET ion sensor. If a homopolymer is present in the template sequence multiple nucleotides will be incorporated in a single flood cycle, and the detected electrical signal will be proportionally higher.

Assembly

Multiple, fragmented sequence reads must be assembled together on the basis of their overlapping areas.

Sequence assembly refers to aligning and merging fragments of a much longer DNA sequence in order to reconstruct the original sequence. This is needed as current DNA sequencing technology cannot read whole genomes as a continuous sequence, but rather reads small pieces of between 20 and 1000 bases, depending on the technology used. Typically the short fragments, called reads, result from shotgun sequencing genomic DNA, or gene transcripts (ESTs).

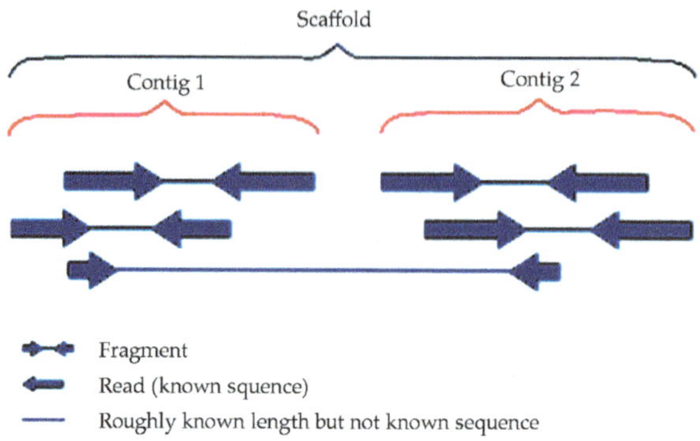

Overlapping reads form contigs; contigs and gaps of known length form scaffolds.

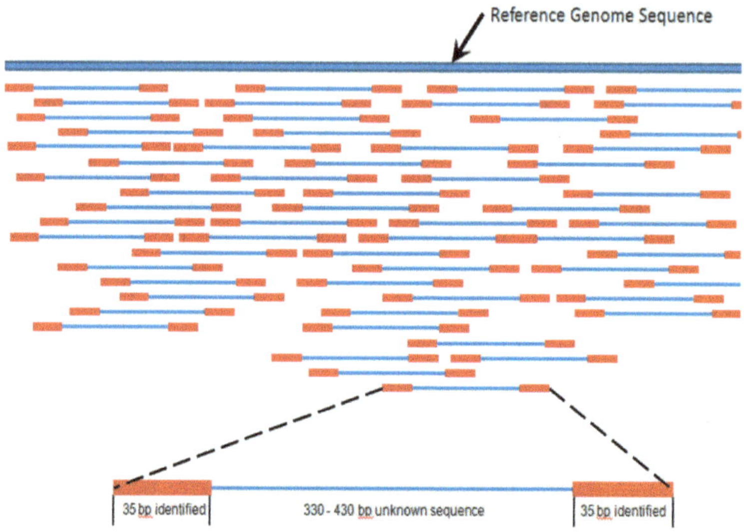

Paired end reads of next generation sequencing data mapped to a reference genome.

Assembly Approaches

Assembly can be broadly categorized into two approaches: *de novo* assembly, for genomes which are not similar to any sequenced in the past, and comparative assembly, which uses the existing sequence of a closely related organism as a reference during assembly. Relative to comparative assembly, *de novo* assembly is computationally difficult (NP-hard), making it less favorable for short-read NGS technologies.

Finishing

Finished genomes are defined as having a single contiguous sequence with no ambiguities representing each replicon.

Annotation

The DNA sequence assembly alone is of little value without additional analysis. **Genome annotation** is the process of attaching biological information to sequences, and consists of three main steps:

- identifying portions of the genome that do not code for proteins
- identifying elements on the genome, a process called gene prediction, and
- attaching biological information to these elements.

Automatic annotation tools try to perform these steps *in silico*, as opposed to manual annotation (a.k.a. curation) which involves human expertise and potential experimental verification. Ideally, these approaches co-exist and complement each other in the same annotation pipeline (also see below).

Traditionally, the basic level of annotation is using BLAST for finding similarities, and then annotating genomes based on homologues. More recently, additional information is added to the annotation platform. The additional information allows manual annotators to deconvolute discrepancies between genes that are given the same annotation. Some databases use genome context information, similarity scores, experimental data, and integrations of other resources to provide genome annotations through their Subsystems approach. Other databases (e.g. Ensembl) rely on both curated data sources as well as a range of software tools in their automated genome annotation pipeline. *Structural annotation* consists of the identification of genomic elements, primarily ORFs and their localisation, or gene structure. *Functional annotation* consists of attaching biological information to genomic elements.

Sequencing Pipelines and Databases

The need for reproducibility and efficient management of the large amount of data associated with genome projects mean that computational pipelines have important applications in genomics.

FUNCTIONAL GENOMICS

Functional genomics is a field of molecular biology that attempts to make use of the vast wealth of data produced by genomic and transcriptomic projects (such as genome sequencing projects and RNA-seq) to describe gene (and protein) functions and interactions. Unlike genomics, functional genomics focuses on the dynamic aspects such as gene transcription, translation, regulation of gene expression and protein–protein interactions, as opposed to the static aspects of the genomic information such as DNA sequence or structures. Functional genomics attempts to answer questions about the function of DNA at the levels of genes, RNA transcripts, and protein products. A key characteristic of functional genomics studies is their genome-wide approach to these questions, generally involving high-throughput methods rather than a more traditional "gene-by-gene" approach.

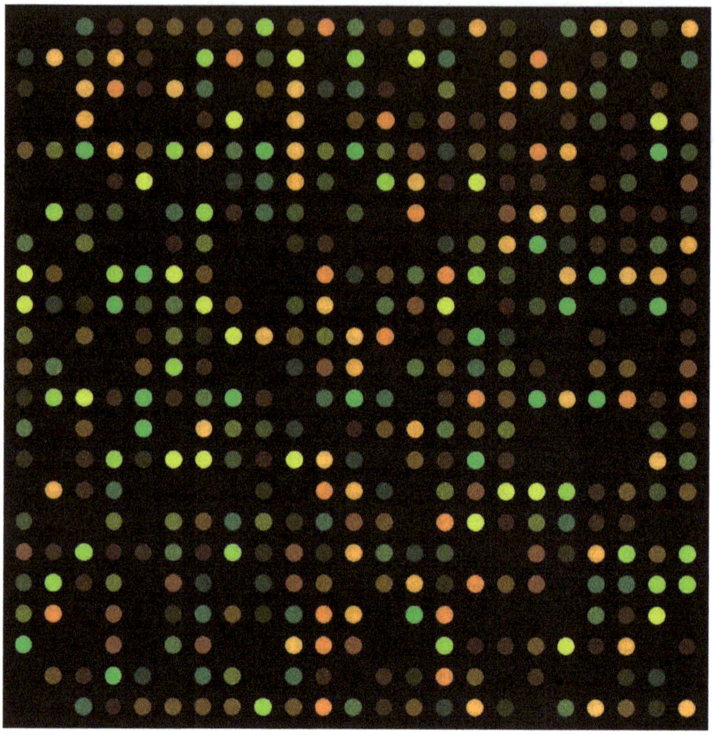

A DNA microarray

Goals of Functional Genomics

The goal of functional genomics is to understand the relationship between an organism's genome and its phenotype. The term functional genomics is often used broadly to refer to the many possible approaches to understanding the properties and function of the entirety of an organism's genes and gene products. This definition is somewhat variable; Gibson and Muse define it as "approaches under development to ascertain the biochemical, cellular, and/or physiological properties of each and every gene product", while Pevsner includes the study of nongenic elements in his definition: "the genome-wide study of the function of DNA (including genes and nongenic elements), as well as the nucleic acid and protein products encoded by DNA". Functional genomics involves studies of natural variation in genes, RNA, and proteins over time (such as an organism's development) or space (such as its body regions), as well as studies of natural or experimental functional disruptions affecting genes, chromosomes, RNA, or proteins.

The promise of functional genomics is to expand and synthesize genomic and proteomic knowledge into an understanding of the dynamic properties of an organism at cellular and/or organismal levels. This would provide a more complete picture of how biological function arises from the information encoded in an organism's genome. The possibility of understanding how a particular mutation leads to a given phenotype has important implications for human genetic diseases, as answering these questions could point scientists in the direction of a treatment or cure.

Techniques and Applications

Functional genomics includes function-related aspects of the genome itself such as mutation and polymorphism (such as single nucleotide polymorphism (SNP) analysis), as well as measurement of molecular activities. The latter comprise a number of "-omics" such as transcriptomics (gene expression), proteomics (protein production), and metabolomics. Functional genomics uses mostly multiplex techniques to measure the abundance of many or all gene products such as mRNAs or proteins within a biological sample. Together these measurement modalities endeavor to quantitate the various biological processes and improve our understanding of gene and protein functions and interactions.

At the DNA Level

Genetic Interaction Mapping

Systematic pairwise deletion of genes or inhibition of gene expression can be used to identify genes with related function, even if they do not interact physically. Epistasis refers to the fact that effects for two different gene knockouts may not be additive; that is, the phenotype that results when two genes are inhibited may be different from the sum of the effects of single knockouts.

The ENCODE Project

The ENCODE (Encyclopedia of DNA elements) project is an in-depth analysis of the human genome whose goal is to identify all the functional elements of genomic DNA, in both coding and noncoding regions. To this point, only the pilot phase of the study has been completed, involving hundreds of assays performed on 44 regions of known or unknown function comprising 1% of the human genome. Important results include evidence from genomic tiling arrays that most nucleotides are transcribed as coding transcripts, noncoding RNAs, or random transcripts, the discovery of additional transcriptional regulatory sites, further elucidation of chromatin-modifying mechanisms.

At the RNA Level: Transcriptome Profiling

Microarrays

Microarrays measure the amount of mRNA in a sample that corresponds to a given gene or probe DNA sequence. Probe sequences are immobilized on a solid surface and allowed to hybridize with fluorescently labeled "target" mRNA. The intensity of fluorescence of a spot is proportional to the amount of target sequence that has hybridized to that spot, and therefore to the abundance of that mRNA sequence in the sample. Microarrays allow for identification of candidate genes involved in a given process based on variation between transcript levels for different conditions and shared expression patterns with genes of known function.

SAGE

SAGE (serial analysis of gene expression) is an alternate method of gene expression analysis based on RNA sequencing rather than hybridization. SAGE relies on the sequencing of 10–17 base pair tags which are unique to each gene. These tags are produced from poly-A mRNA and ligated end-to-end before sequencing. SAGE gives an unbiased measurement of the number of transcripts per cell, since it does not depend on prior knowledge of what transcripts to study (as microarrays do).

Small RNA Sequencing

Small RNAs are a class of non-coding RNA molecules that are key regulators of transcriptional and post-transcriptional gene silencing, or RNA silencing. Next generation sequencing is the gold standard tool for non-coding RNA discovery, profiling and expression analysis.

At the Protein Level: Protein–Protein Interactions

Yeast Two-Hybrid System

A yeast two-hybrid (Y2H) screen tests a "bait" protein against many potential interacting proteins ("prey") to identify physical protein–protein interactions. This system is based on a transcription factor, originally GAL4, whose separate DNA-binding and transcription activation domains are both required in order for the protein to cause transcription of a reporter gene. In a Y2H screen, the "bait" protein is fused to the binding domain of GAL4, and a library of potential "prey" (interacting) proteins is recombinantly expressed in a vector with the activation domain. In vivo interaction of bait and prey proteins in a yeast cell brings the activation and binding domains of GAL4 close enough together to result in expression of a reporter gene. It is also possible to systematically test a library of bait proteins against a library of prey proteins to identify all possible interactions in a cell.

AP/MS

Affinity purification and mass spectrometry (AP/MS) is able to identify proteins that interact with one another in complexes. Complexes of proteins are allowed to form around a particular "bait" protein. The bait protein is identified using an antibody or a recombinant tag which allows it to be extracted along with any proteins that have formed a complex with it. The proteins are then digested into short peptide fragments and mass spectrometry is used to identify the proteins based on the mass-to-charge ratios of those fragments.

Loss-of-Function Techniques

Mutagenesis

Gene function can be investigated by systematically "knocking out" genes one by one. This is done by either deletion or disruption of function (such as by insertional mutagenesis) and the resulting organisms are screened for phenotypes that provide clues to the function of the disrupted gene.

RNAi

RNA interference (RNAi) methods can be used to transiently silence or knock down gene expression using ~20 base-pair double-stranded RNA typically delivered by transfection of synthetic ~20-mer short-interfering RNA molecules (siRNAs) or by virally encoded short-hairpin RNAs (shRNAs). RNAi screens, typically performed in cell culture-based assays or experimental organisms (such as C. elegans) can be used to systematically disrupt nearly every gene in a genome or subsets of genes (sub-genomes); possible functions of disrupted genes can be assigned based on observed phenotypes.

Functional Annotations for Genes

Genome Annotation

Putative genes can be identified by scanning a genome for regions likely to encode proteins, based on characteristics such as long open reading frames, transcriptional initiation sequences, and polyadenylation sites. A sequence identified as a putative gene must be confirmed by further evidence, such as similarity to cDNA or EST sequences from the same organism, similarity of the predicted protein sequence to known proteins, association with promoter sequences, or evidence that mutating the sequence produces an observable phenotype.

Rosetta Stone Approach

The Rosetta stone approach is a computation method of de novo protein function prediction, based on the hypothesis that some proteins involved in a given physiological process may exist as two separate genes in one organism and as a single gene in another. Genomes are scanned for sequences that are independent in one organism and in a single open reading frame in another. If two genes have fused, it is predicted that they have similar biological functions that make such coregulation advantageous.

Functional Genomics and Bioinformatics

Because of the large quantity of data produced by these techniques and the desire to find biologically meaningful patterns, bioinformatics is crucial to analysis of functional genomics data. Examples of techniques in this class are data clustering or principal component analysis for unsupervised machine learning (class detection) as well as artificial neural networks or support vector machines for supervised machine learning (class prediction, classification). Functional enrichment analysis is used to determine the extent of over- or under-expression (positive- or negative- regulators in case of RNAi screens) of functional categories relative to a background sets. Gene ontology based enrichment analysis are provided by DAVID and Gene Set Enrichment Analysis (GSEA), pathway based analysis by Ingenuity and Pathway studio and protein complex based analysis by COMPLEAT.

Structural Genomics

Structural genomics seeks to describe the 3-dimensional structure of every protein encoded by a given genome. This genome-based approach allows for a high-throughput method of structure determination by a combination of experimental and modeling approaches. The principal difference between structural genomics and traditional structural prediction is that structural genomics attempts to determine the structure of every protein encoded by the genome, rather than focusing on one particular protein. With full-genome sequences available, structure prediction can be done more quickly through a combination of experimental and modeling approaches, especially because the availability of large number of sequenced genomes and previously solved protein structures allows scientists to model protein structure on the structures of previously solved homologs.

Because protein structure is closely linked with protein function, the structural genomics has the potential to inform knowledge of protein function. In addition to elucidating protein functions, structural genomics can be used to identify novel protein folds and potential targets for drug discovery. Structural genomics involves taking a large number of approaches to structure determination, including experimental methods using genomic sequences or modeling-based approaches based on sequence or structural homology to a protein of known structure or based on chemical and physical principles for a protein with no homology to any known structure.

As opposed to traditional structural biology, the determination of a protein structure through a structural genomics effort often (but not always) comes before anything is known regarding the protein function. This raises new challenges in structural bioinformatics, i.e. determining protein function from its 3D structure.

Structural genomics emphasizes high throughput determination of protein structures. This is performed in dedicated centers of structural genomics.

While most structural biologists pursue structures of individual proteins or protein groups, specialists in structural genomics pursue structures of proteins on a genome wide scale. This implies large-scale cloning, expression and purification. One main advantage of this approach is economy of scale. On the other hand, the scientific value of some resultant structures is at times questioned. A *Science* article from January 2006 analyzes the structural genomics field.

One advantage of structural genomics, such as the Protein Structure Initiative, is that the scientific community gets immediate access to new structures, as well as to reagents such as clones and protein. A disadvantage is that many of these structures are of proteins of unknown function and do not have corresponding publications. This requires new ways of communicating this structural information to the broader research community. The Bioinformatics core of the Joint center for structural genomics (JCSG) has recently developed a wiki-based approach namely Open protein structure annotation network (TOPSAN) (link) for annotating protein structures emerging from high-throughput structural genomics centers.

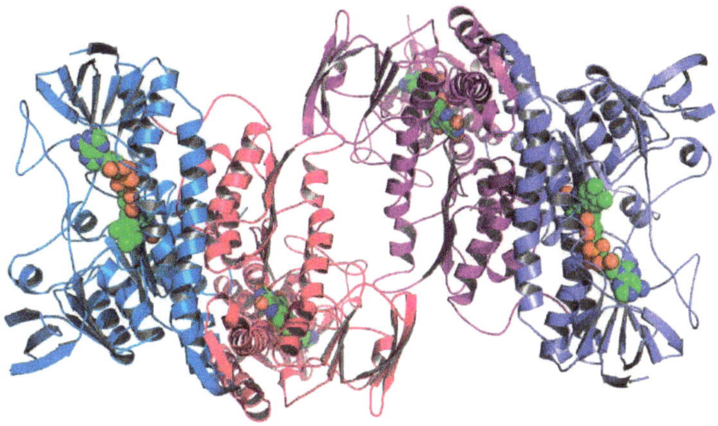

An example of a protein structure from Protein Data Bank.

Goals

One goal of structural genomics is to identify novel protein folds. Experimental methods of protein structure determination require proteins that express and/or crystallize well, which may inherently bias the kinds of proteins folds that this experimental data elucidate. A genomic, modeling-based approach such as *ab initio* modeling may be better able to identify novel protein folds than the experimental approaches because they are not limited by experimental constraints.

Protein function depends on 3-D structure and these 3-D structures are more highly conserved than sequences. Thus, the high-throughput structure determination methods of structural genomics have the potential to inform our understanding of protein functions. This also has potential implications for drug discovery and protein engineering. Furthermore, every protein that is added to the structural database increases the likelihood that the database will include homologous sequences of other unknown proteins. The Protein Structure Initiative (PSI) is a multifaceted effort funded by the National Institutes of Health with various academic and industrial partners that aims to increase knowledge of protein structure using a structural genomics approach and to improve structure-determination methodology.

Methods

Structural genomics takes advantage of completed genome sequences in several ways in order to determine protein structures. The gene sequence of the target protein can also be compared to a known sequence and structural information can then be inferred from the known protein's structure. Structural genomics can be used to predict novel protein folds based on other structural data. Structural genomics can also take modeling-based approach that relies on homology between the unknown protein and a solved protein structure.

De novo Methods

Completed genome sequences allow every open reading frame (ORF), the part of a gene that is likely to contain the sequence for the mRNA and protein, to be cloned and expressed as protein. These proteins are then purified and crystallized, and then subjected to one of two types of structure determination: X-ray crystallography and Nuclear Magnetic Resonance (NMR). The whole genome sequence allows for the design of every primer required in order to amplify all of the ORFs, clone them into bacteria, and then express them. By using a whole-genome approach to this traditional method of protein structure determination, all of the proteins encoded by the genome can be expressed at once. This approach allows for the structural determination of every protein that is encoded by the genome.

Modelling-Based Methods

Ab Initio Modeling

This approach uses protein sequence data and the chemical and physical interactions of the encoded amino acids to predict the 3-D structures of proteins with no homology to solved protein structures. One highly successful method for *ab initio* modeling is the Rosetta program, which divides the protein into short segments and arranges short polypeptide chain into a low-energy local conformation. Rosetta is available for commercial use and for non-commercial use through its public program, Robetta.

Sequence-Based Modeling

This modeling technique compares the gene sequence of an unknown protein with sequences of proteins with known structures. Depending on the degree of similarity between the sequences, the structure of the known protein can be used as a model for solving the structure of the unknown protein. Highly accurate modeling is considered to require at least 50% amino acid sequence identity between the unknown protein and the solved structure. 30-50% sequence identity gives a model of intermediate-accuracy, and sequence identity below 30% gives low-accuracy models. It has been predicted that at least 16,000 protein structures will need to be determined in order for all structural motifs to be represented at least once and thus allowing the structure of any unknown protein to be solved accurately through modeling. One disadvantage of this method, however, is that structure is more conserved than sequence and thus sequence-based modeling may not be the most accurate way to predict protein structures.

Threading

Threading bases structural modeling on fold similarities rather than sequence identity. This method may help identify distantly related proteins and can be used to infer molecular functions.

Examples of Structural Genomics

There are currently a number of on-going efforts to solve the structures for every protein in a given proteome.

The *Thermotogo Maritima* Proteome

One current goal of the Joint Center for Structural Genomics (JCSG), a part of the Protein Structure Initiative (PSI) is to solve the structures for all the proteins in *Thermotogo maritima*, a thermophillic bacterium. *T. maritima* was selected as a structural genomics target based on its relatively small genome consisting of 1,877 genes and the hypothesis that the proteins expressed by a thermophilic bacterium would be easier to crystallize.

Lesley *et al* used *Escherichia coli* to express all the open-reading frames (ORFs) of *T. martima*. These proteins were then crystallized and structures were determined for successfully crystallized proteins using X-ray crystallography. Among other structures, this structural genomics approach allowed for the determination of the structure of the TM0449 protein, which was found to exhibit a novel fold as it did not share structural homology with any known protein.

The *Mycobacterium Tuberculosis* Proteome

The goal of the TB Structural Genomics Consortium is to determine the structures of potential drug targets in *Mycobacterium tuberculosis*, the bacterium that causes tuberculosis. The development of novel drug therapies against tuberculosis are particularly important given the growing problem of multi-drug-resistant tuberculosis.

The fully sequenced genome of *M. tuberculosis* has allowed scientists to clone many of these protein targets into expression vectors for purification and structure determination by X-ray crystallography. Studies have identified a number of target proteins for structure determination, including extracellular proteins that may be involved in pathogenesis, iron-regulatory proteins, current drug targets, and proteins predicted to have novel folds. So far, structures have been determined for 708 of the proteins encoded by *M. tuberculosis*.

Protein Structure Databases and Classifications

- Protein Data Bank (PDB): repository for protein sequence and structural information

- UniProt: provides sequence and functional information
- Structural Classification of Proteins (SCOP Classifications): hierarchical-based approach
- Class, Architecture, Topology and Homologous superfamily (CATH): hierarchical-based approach

EPIGENOMICS

Epigenomics is the study of the complete set of epigenetic modifications on the genetic material of a cell, known as the epigenome. The field is analogous to genomics and proteomics, which are the study of the genome and proteome of a cell. Epigenetic modifications are reversible modifications on a cell's DNA or histones that affect gene expression without altering the DNA sequence. Epigenomic maintenance is a continuous process and plays an important role in stability of eukaryotic genomes by taking part in crucial biological mechanisms like DNA repair. Plant flavones are said to be inhibiting epigenomic marks that cause cancers. Two of the most characterized epigenetic modifications are DNA methylation and histone modification. Epigenetic modifications play an important role in gene expression and regulation, and are involved in numerous cellular processes such as in differentiation/development and tumorigenesis. The study of epigenetics on a global level has been made possible only recently through the adaptation of genomic high-throughput assays.

Introduction to Epigenetics

The mechanisms governing phenotypic plasticity, or the capacity of a cell to change its state in response to stimuli, have long been the subject of research (Phenotypic plasticity 1). The traditional central dogma of biology states that the DNA of a cell is transcribed to RNA, which is translated to proteins, which perform cellular processes and functions. A paradox exists, however, in that cells exhibit diverse responses to varying stimuli and that cells sharing identical sets of DNA such as in multicellular organisms can have a variety of distinct functions and phenotypes. Classical views have attributed phenotypic variation to differences in primary DNA structure, be it through aberrant mutation or an inherited sequence allele. However, while this did explain some aspects of variation, it does not explain how tightly coordinated and regulated cellular responses, such as differentiation, are carried out.

A more likely source of cellular plasticity is through the Regulation of gene expression, such that while two cells may have near identical DNA, the differential expression of certain genes results in variation. Research has shown that cells are capable of regulating gene expression at several stages: mRNA transcription, processing and transportation as well as in protein translation, post-translational processing and degradation. Regulatory proteins that bind to DNA, RNA, and/or proteins are key effectors in these processes and function by positively or negatively regulating specific protein level and function in a cell. And, while DNA binding transcription factors provide a mechanism for specific control of cellular responses, a model where DNA binding transcription factors are the sole regulators of gene activity is also unlikely. For example, in a study of Somatic-cell nuclear transfer, it was demonstrated that stable features of differentiation remain after the nucleus is transferred to a new cellular environment, suggesting that a stable and heritable mechanism of gene regulation was involved in the maintenance of the differentiated state in the absence of the DNA binding transcription factors.

With the finding that DNA methylation and histone modifications are stable, heritable, and also reversible processes that influence gene expression without altering DNA primary structure, a mechanism for the observed variability in cell gene expression was provided. These modifications were termed epigenetic, from epi "on top of" the genetic material "DNA" (Epigenetics 1). The mechanisms governing epigenetic modifications are complex, but through the advent of high-throughput sequencing technology they are now becoming better understood.

Epigenetics

Genomic modifications that alter gene expression that cannot be attributed to modification of the primary DNA sequence and that are heritable mitotically and meiotically are classified as epigenetic modifications. DNA methylation and histone modification are among the best characterized epigenetic processes.

DNA Methylation

The first epigenetic modification to be characterized in depth was DNA methylation. As its name implies, DNA methylation is the process by which a methyl group is added to DNA. The enzymes responsible for catalyzing this reaction are the DNA methyltransferases (DNMTs). While DNA methylation is stable and heritable, it can be reversed by an antagonistic group of enzymes known as DNA de-methylases. In eukaryotes, methylation is most commonly found on the carbon 5 position of cytosine residues (5mC) adjacent to guanine, termed CpG dinucleotides.

DNA methylation patterns vary greatly between species and even within the same organism. The usage of methylation among animals is quite different; with vertebrates exhibiting the highest levels of 5mC and invertebrates more moderate levels of 5mC. Some organisms like *Caenorhabditis elegans* have not been demonstrated to have 5mC nor a conventional DNA methyltransferase; this would suggest that other mechanisms other than DNA methylation are also involved.

Within an organism, DNA methylation levels can also vary throughout development and by region. For example, in mouse primordial germ cells, a genome wide de-methylation even occurs; by implantation stage, methylation levels return to their prior somatic levels. When DNA methylation occurs at promoter regions, the sites of transcription initiation, it has the effect of repressing gene expression. This is in contrast to unmethylated promoter regions which are associated with actively expressed genes.

The mechanism by which DNA methylation represses gene expression is a multi-step process. The distinction between methylated and unmethylated cytosine residues is carried out by specific DNA-binding proteins. Binding of these proteins recruit histone deacetylases (HDACs) enzyme which initiate chromatin remodeling such that the DNA becoming less accessible to transcriptional machinery, such as RNA polymerase, effectively repressing gene expression.

Histone Modification

In eukaryotes, genomic DNA is coiled into protein-DNA complexes called chromatin. Histones, which are the most prevalent type of protein found in chromatin, function to condense the DNA; the net positive charge on histones facilitates their bonding with DNA, which is negatively charged. The basic and repeating units of chromatin, nucleosomes, consist of an octamer of histone proteins (H2A, H2B, H3 and H4) and a 146 bp length of DNA wrapped around it. Nucleosomes and the DNA connecting form a 10 nm diameter chromatin fiber, which can be further condensed.

Chromatin packaging of DNA varies depending on the cell cycle stage and by local DNA region. The degree to which chromatin is condensed is associated with a certain transcriptional state. Unpackaged or loose chromatin is more transcriptionally active than tightly packaged chromatin because it is more accessible to transcriptional machinery. By remodeling chromatin structure and changing the density of DNA packaging, gene expression can thus be modulated.

Chromatin remodeling occurs via post-translational modifications of the N-terminal tails of core histone proteins. The collective set of histone modifications in a given cell is known as the histone code. Many different types of histone modification are known, including: acetylation, methylation, phosphorylation, ubiquitination, SUMOylation, ADP-ribosylation, deamination and proline isomerization; acetylation, methylation, phosphorylation and ubiquitination have been implicated in gene activation whereas methylation, ubiquitination, SUMOylation, deamination and proline isomerization have been implicated in gene repression. Note that several modification types including methylation, phosphorylation and ubiquitination can be associated with different transcriptional states depending on the specific amino acid on the histone being modified. Furthermore, the DNA region where histone modification occurs can also elicit different effects; an example being methylation of the 3rd core histone at lysine residue 36 (H3K36). When H3K36 occurs in the coding sections of a gene, it is associated with gene activation but the opposite is found when it is within the promoter region.

Histone modifications regulate gene expression by two mechanisms: by disruption of the contact between nucleosomes and by recruiting chromatin remodeling ATPases. An example of the first mechanism occurs during the acetylation of lysine terminal tail amino acids, which is catalyzed by histone acetyltransferases (HATs). HATs are part of a multiprotein complex that is recruited to chromatin when activators bind to DNA binding sites. Acetylation effectively neutralizes the basic charge on lysine, which was involved in stabilizing chromatin through its affinity for negatively charged DNA. Acetylated histones therefore favor the dissociation of nucleosomes and thus unwinding of chromatin can occur. Under a loose chromatin state, DNA is more accessible to transcriptional machinery and thus expression is activated. The process can be reversed through removal of histone acetyl groups by deacetylases.

The second process involves the recruitment of chromatin remodeling complexes by the binding of activator molecules to corresponding enhancer regions. The nucleosome remodeling complexes reposition nucleosomes by several mechanisms, enabling or disabling accessibility of transcriptional machinery to DNA. The SWI/SNF protein complex in yeast is one example of a chromatin remodeling complex that regulates the expression of many genes through chromatin remodeling.

Relation to other Genomic Fields

Epigenomics shares many commonalities with other genomics fields, in both methodology and in its abstract purpose. Epigenomics seeks to identify and characterize epigenetic modifications on a global level, similar to the study of the complete set of DNA in genomics or the complete set of proteins in a cell in proteomics. The logic behind performing epigenetic analysis on a global level is that inferences can be made about epigenetic modifications, which might not otherwise be possible through analysis of specific loci. As in the other genomics fields, epigenomics relies heavily on bioinformatics, which combines the disciplines of biology, mathematics and computer science. However while epigenetic modifications had been known and studied for decades, it is through these advancements in bioinformatics technology that have allowed analyses on a global scale. Many current techniques still draw on older methods, often adapting them to genomic assays as is described in the next section.

Epigenomics Methods

Histone Modification Assays

The cellular processes of transcription, DNA replication and DNA repair involve the interaction between genomic DNA and nuclear proteins. It had been known that certain regions within chromatin were extremely susceptible to DNAse I digestion, which cleaves DNA in a low sequence specificity manner. Such hypersensitive sites were thought to be transcriptionally active regions, as evidenced by their association with RNA polymerase and topoisomerases I and II.

It is now known that sensitivity to DNAse I regions correspond to regions of chromatin with loose DNA-histone association. Hypersensitive sites most often represent promoters regions, which require for DNA to be accessible for DNA binding transcriptional machinery to function.

ChIP-Chip and ChIP-Seq

Histone modification was first detected on a genome wide level through the coupling of chromatin immunoprecipitation (ChIP) technology with DNA microarrays, termed ChIP-Chip. However instead of isolating a DNA-binding transcription factor or enhancer protein through chromatin immunoprecipitation, the proteins of interest are the modified histones themselves. First, histones are cross-linked to DNA in vivo through light chemical treatment (e.g., formaldehyde). The cells are next lysed, allowing for the chromatin to be extracted and fragmented, either by sonication or treatment with a non-specific restriction enzyme (e.g., micrococcal nuclease). Modification-specific antibodies in turn, are used to immunoprecipitate the DNA-histone complexes. Following immunoprecipitation, the DNA is purified from the histones, amplified via PCR and labeled with a fluorescent tag (e.g., Cy5, Cy3). The final step involves hybridization of labeled DNA, both immunoprecipitated DNA and non-immunoprecipitated onto a microarray containing immobilized gDNA. Analysis of the relative signal intensity allows the sites of histone modification to be determined.

ChIP-chip was used extensively to characterize the global histone modification patterns of yeast. From these studies, inferences on the function of histone modifications were made; that transcriptional activation or repression was associated with certain histone modifications and by region. While this method was effective providing near full coverage of the yeast epigenome, its use in larger genomes such as humans is limited.

In order to study histone modifications on a truly genome level, other high-throughput methods were coupled with the chromatin immunoprecipitation, namely: SAGE: serial analysis of gene expression (ChIP-SAGE), PET: paired end ditag sequencing (ChIP-PET) and more recently, next-generation sequencing (ChIP-Seq). ChIP-seq follows the same protocol for chromatin immunoprecipitation but instead of amplification of purified DNA and hybridization to a microarray, the DNA fragments are directly sequenced using next generation parallel re-sequencing. It has proven to be an effective method for analyzing the global histone modification patterns and protein target sites, providing higher resolution than previous methods.

DNA Methylation Assays

Techniques for characterizing primary DNA sequences could not be directly applied to methylation assays. For example, when DNA was amplified in PCR or bacterial cloning techniques, the methylation pattern was not copied and thus the information lost. The DNA hybridization technique used in DNA assays, in which radioactive probes were used to map and identify DNA sequences, could not be used to distinguish between methylated and non-methylated DNA.

Restriction Endonuclease Based Methods

Non Genome-Wide Approaches

The earliest methylation detection assays used methylation modification sensitive restriction endonucleases. Genomic DNA was digested with both methylation-sensitive and insensitive restriction enzymes recognizing the same restriction site. The idea being that whenever the site was methylated, only the methylation insensitive enzyme could cleave at that position. By comparing restriction fragment sizes generated from the methylation-sensitive enzyme to those of the methylation-insensitive enzyme, it was possible to determine the methylation pattern of the region. This analysis step was done by amplifying the restriction fragments via PCR, separating them through gel electrophoresis and analyzing them via southern blot with probes for the region of interest.

This technique was used to compare the DNA methylation modification patterns in the human adult and hemoglobin gene loci. Different regions of the gene (gamma delta beta globin) were known to be expressed at different stages of development. Consistent with a role of DNA methylation in gene repression, regions that were associated with high levels of DNA methylation were not actively expressed.

This method was limited not suitable for studies on the global methylation pattern, or 'methylome'. Even within specific loci it was not fully representative of the true methylation pattern as only those restriction sites with corresponding methylation sensitive and insensitive restriction assays could provide useful information. Further complications could arise when incomplete digestion of DNA by restriction enzymes generated false negative results.

Genome Wide Approaches

DNA methylation profiling on a large scale was first made possible through the Restriction Landmark Genome Scanning (RLGS) technique. Like the locus-specific DNA methylation assay, the technique identified methylated DNA via its digestion methylation sensitive enzymes. However it was the use of two-dimensional gel electrophoresis that allowed be characterized on a broader scale.

However it was not until the advent of microarray and next generation sequencing technology when truly high resolution and genome-wide DNA methylation became possible. As with RLGS, the endonuclease component is retained in the method but it is coupled to new technologies. One such approach is the differential methylation hybridization (DMH), in which one set of genomic DNA is digested with methylation-sensitive restriction enzymes and a parallel set of DNA is not digested. Both sets of DNA are subsequently amplified and each labelled with fluorescent dyes and used in two-colour array hybridization. The level of DNA methylation at a given loci is determined by the relative intensity ratios of the two dyes. Adaptation of next generation sequencing to DNA methylation assay provides several advantages over array hybridization. Sequence-based technology provides higher resolution to allele specific DNA methylation, can be performed on larger genomes, and does not require creation of DNA microarrays which require adjustments based on CpG density to properly function.

Bisulfite Sequencing

Bisulfite sequencing relies on chemical conversion of unmethylated cytosines exclusively, such that they can be identified through standard DNA sequencing techniques. Sodium bisulfate and alkaline treatment does this by converting unmethylated cytosine residues into uracil while leaving methylated cytosine unaltered. Subsequent amplification and sequencing of untreated DNA and sodium bisulphite treated DNA allows for methylated sites to be identified. Bisulfite sequencing, like the traditional restriction based methods, was historically limited to methylation patterns of specific gene loci, until whole genome sequencing technologies became available. However, unlike traditional restriction based methods, bisulfite sequencing provided resolution on a nucleotide level.

Limitations of the bisulfite technique include the incomplete conversion of cytosine to uracil, which is a source of false positives. Further, bisulfite treatment also causes DNA degradation and requires an additional purification step to remove the sodium bisulfite.

Next-generation sequencing is well suited in complementing bisulfite sequencing in genome-wide methylation analysis. While this now allows for methylation pattern to be determined on the highest resolution possible, on the single nucleotide level, challenges still remain in the assembly step because of reduced sequence complexity in bisulphite treated DNA. Increases in read length seek to address this challenge, allowing for whole genome shotgun bisulphite sequencing (WGBS) to be performed. The WGBS approach using an Illumina Genome Analyzer platform and has already been implemented in *Arabidopsis thaliana*.

Direct Detection

Polymerase sensitivity in single molecule real time sequencing made it possible for scientists to directly detect epigenetic marks such as methylation as the polymerase moves along the DNA molecule being sequenced. Several projects have demonstrated the ability to collect genome-wide epigenetic data in bacteria.

Theoretical Modeling Approaches

First mathematical models for different nucleosome states affecting gene expression were introduced in 1980s ref. Later, this idea was almost forgotten, until the experimental evidence has indicated a possible role of covalent histone modifications as an epigenetic code. In the next several years, high-throughput data have indeed uncovered the abundance of epigenetic modifications and their relation to chromatin functioning which has motivated new theoretical models for the appearance, maintaining and changing these patterns,. These models are usually formulated in the frame of one-dimensional lattice approaches.

Metagenomics

Metagenomics is the study of *metagenomes*, genetic material recovered directly from environmental samples. The broad field may also be referred to as environmental genomics, ecogenomics or community genomics. While traditional microbiology and microbial genome sequencing rely upon cultivated clonal cultures, early environmental gene sequencing cloned specific genes (often the 16S rRNA gene) to produce a profile of diversity in a natural sample. Such work revealed that the vast majority of microbial biodiversity had been missed by cultivation-based methods. Recent studies use "shotgun" Sanger sequencing or massively parallel pyrosequencing to get largely unbiased samples of all genes from all the members of the sampled communities. Because of its power to reveal the previously hidden diversity of microscopic life, metagenomics offers a powerful lens for viewing the microbial world that has the potential to revolutionize understanding of the entire living world.

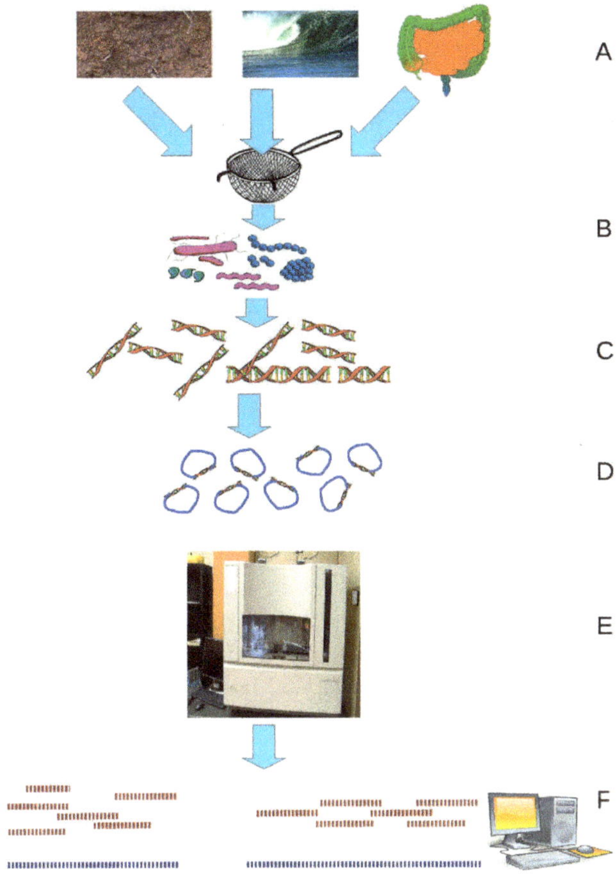

Environmental Shotgun Sequencing (ESS) is a key technique in metagenomics. (A) Sampling from habitat; (B) filtering particles, typically by size; (C) Lysis and DNA extraction; (D) cloning and library construction; (E) sequencing the clones; (F) sequence assembly into contigs and scaffolds.

Study Systems

Viruses and Bacteriophages

Bacteriophages have played and continue to play a key role in bacterial genetics and molecular biology. Historically, they were used to define gene structure and gene regulation. Also the first genome to be sequenced was a bacteriophage. However, bacteriophage research did not lead the genomics revolution, which is clearly dominated by bacterial genomics. Only very recently has the study of bacteriophage genomes become prominent, thereby enabling researchers to understand the mechanisms underlying phage evolution. Bacteriophage genome sequences can be obtained through direct sequencing of isolated bacteriophages, but can also be derived as part of microbial genomes. Analysis of bacterial genomes has shown that a substantial amount of microbial DNA consists of prophage sequences and prophage-like elements. A detailed database mining of these sequences offers insights into the role of prophages in shaping the bacterial genome.

Cyanobacteria

At present there are 24 cyanobacteria for which a total genome sequence is available. 15 of these cyanobacteria come from the marine environment. These are six *Prochlorococcus* strains, seven marine *Synechococcus* strains, *Trichodesmium erythraeum* IMS101 and *Crocosphaera watsonii* WH8501. Several studies have demonstrated how these sequences could be used very successfully to infer important ecological and physiological characteristics of marine cyanobacteria. However, there are many more genome projects currently in progress, amongst those there are further *Prochlorococcus* and marine *Synechococcus* isolates, *Acaryochloris* and *Prochloron*, the N_2-fixing filamentous cyanobacteria *Nodularia spumigena*, *Lyngbya aestuarii* and *Lyngbya majuscula*, as well as bacteriophages infecting marine cyanobaceria. Thus, the growing body of genome information can also be tapped in a more general way to address global problems by applying a comparative approach. Some new and exciting examples of progress in this field are the identification of genes for regulatory RNAs, insights into the evolutionary origin of photosynthesis, or estimation of the contribution of horizontal gene transfer to the genomes that have been analyzed.

Applications of Genomics

Genomics has provided applications in many fields, including medicine, biotechnology, anthropology and other social sciences.

Genomic Medicine

Next-generation genomic technologies allow clinicians and biomedical researchers to drastically increase the amount of genomic data collected on large study populations. When combined with new informatics approaches that integrate many kinds of data with genomic data in disease research, this allows researchers to better understand the genetic bases of drug response and disease.

Synthetic Biology and Bioengineering

The growth of genomic knowledge has enabled increasingly sophisticated applications of synthetic biology. In 2010 researchers at the J. Craig Venter Institute announced the creation of a partially synthetic species of bacterium, *Mycoplasma laboratorium*, derived from the genome of *Mycoplasma genitalium*.

Conservation Genomics

Conservationists can use the information gathered by genomic sequencing in order to better evaluate genetic factors key to species conservation, such as the genetic diversity of a population or whether an individual is heterozygous for a recessive inherited genetic disorder. By using genomic data to evaluate the effects of evolutionary processes and to detect patterns in variation throughout a given population, conservationists can formulate plans to aid a given species without as many variables left unknown as those unaddressed by standard genetic approaches.

LIST OF SEQUENCED PLANT GENOMES

This list of sequenced plant genomes contains plant species known to have publicly available complete genome sequences that have been assembled, annotated and published. Unassembled genomes are not included, nor are organelle only sequences. For all kingdoms, see the list of sequenced genomes.

Algae

Unicellular photosynthetic eukaryotes.

Organism strain	**Clade**	**Relevance**	**Genome size**	**Number of genes predicted**	**Organization**	**Year of completion**	**Assembly status**	**Links**
Cyanophora paradoxa	Glaucophyte				Rutgers University	2012		
Bathycoccus prasinos BBAN7	Green algae	Comparative analysis	15 Mb		Joint Genome Institute	2012		
Chlamydomonas reinhardtii CC-503 cw92 mt+	Green algae	Model organism	111 Mb	17,737	University of California at Los Angeles	2007		*"Chlamydomonas reinhardtii". National Center for Biotechnology Information (NCBI).*ENAGCA_000002595
Chlorella variabilis NC64A	Green algae					2010		
*Coccomyxa subellipsoidea*sp. C-169	Green algae	Model biofuel			Joint Genome Institute	2007		
*Dunaliella salina*CCAP19/18	Green algae	Halophilic, biofuel and beta-carotene production			Joint Genome Institute		Organelle genomes complete, nuclear genome in progress	
Micromonas pusilla CCMP1545	Green algae	Marine phytoplankton			Joint Genome Institute	2007		
*Micromonas pusilla*RCC299/ NOUM17	Green algae	Marine phytoplankton			Joint Genome Institute	2007		
Ostreococcus lucimarinus CCE9901	Green algae	Simple eukaryote, small genome	13.2 Mb	7,796		2007		
Ostreococcus tauri OTH95	Green algae	Simple eukaryote, small genome				2006		
Ostreococcus sp. RCC809	Green algae			7,773	Joint Genome Institute	2008		
Volvox carteri	Green algae	Multicellular alga, model organism	~131.2 Mb	14,971		2010		
Chondrus crispus	Red algae		105 Mb	9,606	Genoscope/Station Biologique de Roscoff	2013		

Organism strain	Clade	Relevance	Genome size	Number of genes predicted	Organization	Year of completion	Assembly status	Links
*Cyanidioschyzon merolae*Strain:10D	Red algae	Photo-autotrophic	16.73 Mb	5,017		2004,2007		
Galdieria sulphuraria	Red algae	Thermo-acidophilic (extremophile)	13.7 Mb	6,623		2005 2005 2013		
Porphyridium purpureum	Red algae		19.7 Mb	8,355		2013		
Pyropia yezoensis	Red algae		43 Mb	10,327		2013		
Ectocarpus siliculosus	Brown algae (Heterokontophyta)	distantly related to plants			Station Biologique de Roscoff	2010		

Bryophytes

Organism strain	Division	Relevance	Genome size	Number of genes predicted	Organization	Year of completion	Assembly status
Physcomitrella patens ssp. patens str. Gransden 2004	Bryophytes	Early diverging land plant				2008	

Higher Plants (Vascular Plants)

Organism strain	Division	Relevance	Genome size	Number of genes predicted	Organization	Year of completion	Assembly status
Selaginella moellendorffii	Lycopodiophyta	Model organism				2011	

Angiosperms

Amborellales

Organism strain	Family	Relevance	Genome size	Number of genes predicted	Organization	Year of completion	Assembly status
Amborella trichopoda	Amborellaceae	Basal angiosperm				2013	

Eudicots

Ranunculales

Organism strain	Family	Relevance	Genome size	Number of genes predicted	Organization	Year of completion	Assembly status
Aquilegia caerulea	Ranunculaceae	Basal eudicot				Unpublished	

Proteales

Organism strain	Family	Relevance	Genome size	Number of genes predicted	Organization	Year of completion	Assembly status
Nelumbo nucifera	Nelumbonaceae	Basal eudicot				2013	

Caryophyllales

Organism strain	Family	Relevance	Genome size	Number of genes predicted	Organization	Year of completion	Assembly status
Beta vulgaris (sugar beet)	Chenopodiaceae	Crop plant	714–758 Mbp	27,421		2013	

Rosids

Organism strain	Family	Relevance	Genome size	Number of genes predicted	Organization	Year of completion	Assembly status
Betula nana (dwarf birch)	Betulaceae	Arctic shrub	450 Mbp		QMUL/SBCS	2013	
Aethionema arabicum	Brassicaceae	Comparative analysis of crucifer genomes				2013	
Arabidopsis lyrata	Brassicaceae	model plant				2011	
Arabidopsis thaliana Ecotype: Columbia	Brassicaceae	Model plant	135 Mbp			2000	
Brassica rapa (Chinese cabbage)	Brassicaceae	Assorted crops and model organism				2011	
Capsella rubella	Brassicaceae	Close relative of Arabidopsis thaliana	130Mbp	26,521	JGI	2013?2013	
Eutrema salsugineum	Brassicaceae	A relative of arabidopsis with high salt tolerance	240Mbp	26,351	JGI	2013	
Eutrema parvulum	Brassicaceae	Comparative analysis of crucifer genomes				2013	

Organism strain	Family	Relevance	Genome size	Number of genes predicted	Organization	Year of completion	Assembly status
Leavenworthia alabamica	Brassicaceae	Comparative analysis of crucifer genomes				2013	
Sisymbrium irio	Brassicaceae	Comparative analysis of crucifer genomes				2013	
Thellungiella parvula	Brassicaceae	A relative of arabidopsis with high salt tolerance				2011	
Cannabis sativa (hemp)	Cannabaceae	Hemp and marijuana production	ca 820Mbp	30,074 based on transcriptome assembly and clustering		2011	Illumina/454 scaffold N50 16.2 Kbp
Carica papaya (papaya)	Caricaceae	Fruit crop	372Mbp	28,629		2008	contig N50 11kbp scaffold N50 1Mbp total coverage ~3x (Sanger) 92.1% unigenes mapped 235Mbp anchored (of this 161Mbp also oriented)
Kalanchoe	Crassulaceae					2013?	
Citrullus lanatus (watermelon)	Cucurbitaceae	Vegetable crop	ca 425Mbp	23,440	BGI	2012	Illumina coverage 108.6x contig N50 26.38 kbp Scaffold N50 2.38 Mbp genome covered 83.2% ~97% ESTs mapped
Cucumis melo (Muskmelon) DHL92	Cucurbitaceae	Vegetable crop	450Mbp	27,427		2012	454 13.5x coverage contig N50: 18.1kbp scaffold N50: 4.677 Mbp WGS

Organism strain	**Family**	**Relevance**	**Genome size**	**Number of genes predicted**	**Organization**	**Year of completion**	**Assembly status**
Cucumis sativus (cucumber) 'Chinese long' inbred line 9930	Cucurbitaceae	Vegetable crop	350 Mbp (Kmer depth) 367 Mbp (flow cytometry)	26,682		2009	contig N50 19.8kbp scaffold N50 1,140kbp total coverage ~72.2 (Sanger + Ilumina) 96.8% unigenes mapped 72.8% of the genome anchored
Hevea brasiliensis (rubber tree)	Euphorbiaceae	the most economically important member of the genus Hevea				2013	
*Jatropha curcas*Palawan	Euphorbiaceae	bio-diesel crop				2010	
Manihot esculenta (Cassava)	Euphorbiaceae	Humanitarian importance	~760Mb	30,666	JGI	2012	
Ricinus communis (Castor bean)	Euphorbiaceae	Oilseed crop	320Mbp	31,237	JCVI	2010	Sanger coverage~4.6x contig N50 21.1 kbp scaffold N50 496.5kbp
Cajanus cajan (Pigeon pea) var. Asha	Fabaceae	Model legume				2012	
Arachis duranensis (A genome diploid wild peanut) accession V14167	Fabaceae	Wild ancestor of peanut, an oilseed and grain legume crop				2016	Illumina 154x coverage, contig N50 22 kbp, scaffold N50 948 kbp
*Arachis ipaensis (*B genome diploid wild peanut) accession K30076	Fabaceae	Wild ancestor of peanut, an oilseed and grain legume crop				2016	Illumina 163x coverage, contig N50 23 kbp, scaffold N50 5,343 kbp
Cicer arietinum (chickpea)	Fabaceae	filling				2013	
Cicer arietinum L. (chickpea)	Fabaceae					2013	
Glycine max (soybean) var. Williams 82	Fabaceae	Protein and oil crop	1115Mbp	46,430		2010	Contig N50:189.4kbp Scaffold N50:47.8Mbp Sanger coverage ~8x WGS 955.1 Mbp assembled
Lotus japonicus (Bird's-foot Trefoil)	Fabaceae	Model legume				2008	
Medicago truncatula (Barrel Medic)	Fabaceae	Model legume				2011	
Phaseolus vulgaris (common bean)	Fabaceae	Model bean	520Mbp	31,638	JGI	2013?	
Linum usitatissimum (flax)	Linaceae	Crop	~350Mbp	43,384	BGI et al.	2012	

Organism strain	Family	Relevance	Genome size	Number of genes predicted	Organization	Year of completion	Assembly status
Gossypium raimondii	Malvaceae	One of the putative progenitor species of tetraploid cotton				2013?	
Theobroma cacao (cocoa tree)	Malvaceae	Flavouring crop				2010	
Theobroma cacao (cocoa tree) cv. Matina 1-6	Malvaceae	Most widely cultivated cacao type				2013	
Azadirachta indica (neem)	Meliaceae	Source of number of Terpenoids, including biopesticide azadirachtin, Used in Traditional Medicine	364 Mbp	~20000	GANIT Labs	2012 and 2011	Illumina GAIIx, scaffold N50 of 452028bp, Transcriptome data from Shoot, Root, Leaf, Flower and Seed
Eucalyptus grandis (Rose gum)	Myrtaceae	Fibre and timber crop				2011	
Fragaria vesca (wild strawberry)	Rosaceae	Fruit crop	240Mbp	34,809		2011	scaffold N50: 1.3 Mbp 454/Illumina/ solid 39x coverage WGS
Malus domestica (apple) "Golden Delicious"	Rosaceae	Fruit crop	~742.3Mbp	57,386		2010	contig N50 13.4 (kbp??) scafold N50 1,542.7 (kbp??) total coverage ~16.9x (Sanger + 454) 71.2% anchored
Prunus amygdalus (almond)	Rosaceae	Fruit crop				2013?	
Prunus avium (sweet cherry) cv. Stella	Rosaceae	Fruit crop				2013?	
Prunus mume (Chinese plum or Japanese apricot)	Rosaceae	Fruit crop				2012	
Prunus persica (peach)	Rosaceae	Fruit crop	265Mbp	27,852		2013	Sanger coverage:8.47x WGS ca 99% ESTs mapped 215.9 Mbp in pseudomolecules
Pyrus bretschneideri (ya pear or Chinese white pear) cv. Dangshansuli	Rosaceae	Fruit crop				2012	
Pyrus communis (European pear) cv. Doyenne du Comice	Rosaceae	Fruit crop				2013?	
Citrus clementina (Clementine)	Rutaceae	Fruit crop				2013?	

Organism strain	Family	Relevance	Genome size	Number of genes predicted	Organization	Year of completion	Assembly status
Citrus sinensis (Sweet orange)	Rutaceae	Fruit crop				2013?,2013	
Populus trichocarpa (poplar)	Salicaceae	Carbon sequestration, model tree, timber	510 Mbp (cytogenetic) 485 Mbp (coverage)	73,013 Phytozome		2006	Scaffold N50: 19.5 Mbp Contig N50:552.8 Kbp phytozome WGS >=95 % cDNA found
Vitis vinifera (grape) genotype PN40024	Vitaceae	fruit crop				2007	

Asterids

Organism strain	Family	Relevance	Genome size	Number of genes predicted	Organization	Year of completion	Assembly status
Mimulus guttatus	Scrophulariaceae	model system for studying ecological and evolutionary genetics	ca 430Mbp	26,718	JGI	2013?	Scaffold N50 = 1.1 Mbp Contig N50 = 45.5 Kbp
Solanum lycopersicum (tomato) cv. Heinz 1706	Solanaceae	Food crop	ca 900Mbp	34,727	SGN	20112012	Sanger/454/ Illumina/Solid Pseudomolecules spanning 91 scaffolds (760Mbp of which 594Mbp have been oriented) over 98% ESTs mappable
Solanum pimpinellifolium (Currant Tomato)	Solanaceae	closest wild relative to tomato				2012	Illumina contig N50: 5100bp ~40x coverage
Solanum tuberosum (potato)	Solanaceae	Food crop	844 Mbp kmer (856 Mbp)	39,031	PGSC	2011	Sanger/454/ Illumina 79.2x coverage contig N50: 31,429bp scaffold N50: 1,318,511bp

Organism strain	Family	Relevance	Genome size	Number of genes predicted	Organization	Year of completion	Assembly status
Solanum commersonii (commerson's nightshade)	Solanaceae	Wild potato relative	838 Mbp kmer (840 Mbp)	37,662	UNINA, UMN, UNIVR, Sequentia Biotech, CGR	2015	Illumina 105x coverage contig N50: 6,506bp scaffold N50: 44,298bp
Nicotiana benthamiana	Solanaceae	Close relative of tobacco	ca 3Gbp			2012	Illumina 63x coverage contig N50: 16,480bp scaffold N50:89,778bp >93% unigenes found
Nicotiana sylvestris (Tobacco plant)	Solanaceae	model system for studies of terpenoid production	2.636Gbp		Philip Morris International	2013	94x coverage scaffold N50: 79.7 kbp 194kbp superscaffolds using physical Nicotiana map
Nicotiana tomentosiformis	Solanaceae	Tobacco progenitor	2.682 Gb		Philip Morris International	2013	146x coverage scaffold N50: 82.6 kb 166kbp superscaffolds using physical Nicotiana map
Capsicum annuum(Pepper) (a) cv. CM334 (b) cv. Zunla-1	Solanaceae	Food crop	~3.48 Gbp	(a) 34,903 (b) 35,336		(a) 2014 (b) 2014	N50 contig: (a) 30.0 kb (b) 55.4 kb N50 scaffold: (a) 2.47 Mb (b) 1.23 Mb
Capsicum annuum var. glabriusculum (Chiltepin)	Solanaceae	Progenitor of cultivated pepper	~3.48 Gbp	34,476		2014	N50 contig: 52.2 kb N50 scaffold: 0.45 Mb
Petunia	Solanaceae	Economically important flower				2011	
Utricularia gibba (humped bladderwort)	Lentibulariaceae	model system for studying genome size evolution; a carnivorous plant	81.87 Mb	28,494	LANGEBIO, CINVESTAV	2013	Scaffold N50: 80.839 Kb

Monocots

Grasses

Organism strain	Family	Relevance	Genome size	Number of genes predicted	**Organization**	**Year of completion**	**Assembly status**
Setaria italica (Foxtail millet)	Poaceae	Model of C4 metabolism				2012	
Aegilops tauschii (Tausch's goatgrass)	Poaceae	bread wheat D-genome progenitor	ca 4.36Gb		BGI	2013	Non-repetitive sequence assembled
Brachypodium distachyon (purple false brome)	Poaceae	Model monocot				2010	
Hordeum vulgare (barley)	Poaceae	Model of ecological adoption			IBSC	2012	
Oryza brachyantha (wild rice)	Poaceae	Disease resistant wild relative of rice				2013	
Oryza glaberrima (African rice) var CG14	Poaceae	West-African species of rice				2010	
Oryza rufipogon (red rice)	Poaceae	Ancestor to Oryza sativa	406 Mb	37,071	SIBS	2012	Illumina HiSeq2000 100x coverage
Oryza sativa (long grain rice) ssp indica	Poaceae	Crop and model cereal				2002	
Oryza sativa (Short grain rice) ssp japonica	Poaceae	Crop and model cereal				2002	
Panicum virgatum (switchgrass)	Poaceae	biofuel				2013?	
Phyllostachys edulis (moso bamboo)	Poaceae					2013	
Sorghum bicolor genotype BTx623	Poaceae	Crop	ca 730Mbp	34,496		2009	contig N50:195.4kbp scaffold N50: 62.4Mbp Sanger, 8.5x coverage WGS
Triticum aestivum (bread wheat)	Poaceae	20% of global nutrition				2012	Non-repetitive sequence assembled Roche 454/ Illumina WGS
Triticum urartu	Poaceae	Bread wheat A-genome progenitor	ca 4.94Gb		BGI	2013	Non-repetitive sequence assembled Illumina WGS
Zea mays (maize) ssp *mays*B73	Poaceae	Cereal crop	2,300Mbp	39,656		2009	contig N50 40kbp scaffold N50: 76kbp Sanger, 4-6x coverage per BAC

Other Non-Grasses

Organism strain	Family	Relevance	Genome size	Number of genes predicted	Organization	Year of completion	Assembly status
Musa acuminata (Banana)	Musaceae	A-genome of modern banana cultivars	523 Mbp	36,542		2012	N50 contig: 43.1 kb N50 scaffold: 1.3 Mb
Musa balbisiana (Wild banana)	Musaceae	B-genome of modern banana cultivars	438 Mbp	36,638		2013	N50 contig: 7.9 kb
Phoenix dactylifera (Date palm)	Arecaceae	Woody crop in arid regions	658 Mbp	28,800		2011	N50 contig: 6.4 kb
Elaeis guineensis (African oil palm)	Arecaceae	Oil-bearing crop	~1800 Mbp	34,800		2013	N50 scaffold: 1.27 Mb
Spirodela polyrhiza (Greater duckweed)	Araceae	Aquatic plant	158 Mbp	19,623		2014	N50 scaffold: 3.76 Mb

Gymnosperm

Organism strain	Family	Relevance	Genome size	Number of genes predicted	Organization	Year of completion	Assembly status
Picea abies (Norway spruce)	Pinaceae	Timber, tonewood, ornamental such as Christmas tree	20 Gb	28,354	Umeå Plant Science Centre / SciLifeLab, Sweden	2013	
Picea glauca (White spruce)	Pinaceae	Timber, Pulp	20.8 Gb	56,064	Institutional Collaboration	2013	
Pinus taeda (Loblolly pine)	Pinaceae	Timber	20.15 Gb	50,172	Institutional collaboration	2014	N50 scaffold size: 66.9 kbp

Uncategorised Things to Add...

the genome from Galdieria sulphuraria has finally been published (Schönknecht, G., W.-H. Chen, et al. (2013). "Gene transfer from bacteria and archaea facilitated evolution of an extremophilic eukaryote." Science 339(6124): 1207-1210.) Genome size is 13.7 MB, and 6623 protein-coding genes were annotated.

Nakamura et al. published the genome sequence for Pyropia yezoensis (Nakamura, Y., N. Sasaki, et al. (2013). "The first symbiont-free genome sequence of marine red alga, Susabi-nori Pyropia yezoensis." PLoS ONE 8(3): e57122.).

Bhattacharya et al. published the genome of Porphyridium purpureum (Bhattacharya, D., D. C. Price, et al. (2013). "Genome of the red alga Porphyridium purpureum." Nature Communications 4.)

Press Releases Announcing Sequencing

Not meeting criteria of the first paragraph of this article in being nearly full sequences with high quality, published, assembled and publicly available. This list includes species where sequences are announced in press releases or websites, but not in a data-rich publication in a refereed Journal with doi.

- *Brassica napus*, oil plant (2009)
- *Elaeis guineensis*, oil palm (2007)
- *Corchorus olitorius*, fibre plant (2010)
- *Fraxinus excelsior*, European ash (2013 draft)

REFERENCES

1. National Human Genome Research Institute (2010-11-08). "FAQ About Genetic and Genomic Science". Genome.gov. Retrieved 2011-12-03.
2. National Human Genome Research Institute (2010-11-08). "A Brief Guide to Genomics". Genome.gov. Retrieved 2011-12-03.
3. Concepts of genetics (10th ed.). San Francisco: Pearson Education. 2012. ISBN 9780321724120.
4. Kadakkuzha BM, Puthanveettil SV (Jul 2013). "Genomics and proteomics in solving brain complexity". Molecular bioSystems. 9 (7): 1807–21. doi:10.1039/C3MB25391K. PMID 23615871.
5. Pevsner J (2009). Bioinformatics and functional genomics (2nd ed.). Hoboken, N.J: Wiley-Blackwell. ISBN 9780470085851.
6. Culver KW, Labow MA (2002-11-08). "Genomics". In Robinson R. Genetics. Macmillan Science Library. Macmillan Reference USA. ISBN 0028656067.
7. "Henry George Liddell, Robert Scott, An Intermediate Greek-English Lexicon, γίγνομαι". Retrieved 2015-05-13.
8. "Genome, n". Oxford English Dictionary (Third ed.). Oxford University Press. 2008. Retrieved 2012-12-01.(subscription required)
9. Yadav SP (Dec 2007). "The wholeness in suffix -omics, -omes, and the word om". Journal of Biomolecular Techniques. 18 (5): 277. PMC 2392988 . PMID 18166670.
10. Ankeny RA (Jun 2003). "Sequencing the genome from nematode to human: changing methods, changing science". Endeavour. 27 (2): 87–92. doi:10.1016/S0160-9327(03)00061-9. PMID 12798815.
11. Holley RW, Everett GA, Madison JT, Zamir A (May 1965). "Nucleotide sequences in the yeast alanine transfer ribonucleic acid" (PDF). The Journal of Biological

Chemistry. 240 (5): 2122–8. PMID 14299636.

12. Holley RW, Apgar J, Everett GA, Madison JT, Marquisee M, Merrill SH, Penswick JR, Zamir A (Mar 1965). "Structure of a ribonucleic acid". Science. 147 (3664): 1462–5. Bibcode:1965Sci...147.1462H. doi:10.1126/science.147.3664.1462. PMID 14263761.
13. Nirenberg M, Leder P, Bernfield M, Brimacombe R, Trupin J, Rottman F, O'Neal C (May 1965). "RNA codewords and protein synthesis, VII. On the general nature of the RNA code". Proceedings of the National Academy of Sciences of the United States of America. 53 (5): 1161–8. Bibcode:1965PNAS...53.1161N. doi:10.1073/pnas.53.5.1161. PMC 301388 . PMID 5330357.
14. Min Jou W, Haegeman G, Ysebaert M, Fiers W (May 1972). "Nucleotide sequence of the gene coding for the bacteriophage MS2 coat protein". Nature. 237 (5350): 82–8. Bibcode:1972Natur.237...82J. doi:10.1038/237082a0. PMID 4555447.
15. Fiers W, Contreras R, Duerinck F, Haegeman G, Iserentant D, Merregaert J, et al. (Apr 1976). "Complete nucleotide sequence of bacteriophage MS2 RNA: primary and secondary structure of the replicase gene". Nature. 260 (5551): 500–7. Bibcode:1976Natur.260..500F. doi:10.1038/260500a0. PMID 1264203.
16. Fiers W, Contreras R, Haegemann G, Rogiers R, Van de Voorde A, Van Heuverswyn H, Van Herreweghe J, Volckaert G, Ysebaert M (May 1978). "Complete nucleotide sequence of SV40 DNA". Nature. 273 (5658): 113–20. Bibcode:1978Natur.273..113F. doi:10.1038/273113a0. PMID 205802.
17. Tamarin RH (2004). Principles of genetics (7 ed.). London: McGraw Hill. ISBN 9780071243209.
18. Sanger F (1980). "Nobel lecture: Determination of nucleotide sequences in DNA" (PDF). Nobelprize.org. Retrieved 2010-10-18.
19. Sanger F, Air GM, Barrell BG, Brown NL, Coulson AR, Fiddes CA, Hutchison CA, Slocombe PM, Smith M (Feb 1977). "Nucleotide sequence of bacteriophage phi X174 DNA". Nature. 265 (5596): 687–95. Bibcode:1977Natur.265..687S. doi:10.1038/265687a0. PMID 870828.
20. Kaiser O, Bartels D, Bekel T, Goesmann A, Kespohl S, Pühler A, Meyer F (Dec 2003). "Whole genome shotgun sequencing guided by bioinformatics pipelines--an optimized approach for an established technique" (PDF). Journal of Biotechnology. 106 (2–3): 121–33. doi:10.1016/j.jbiotec.2003.08.008. PMID 14651855.
21. Sanger F, Nicklen S, Coulson AR (Dec 1977). "DNA sequencing with chain-terminating inhibitors". Proceedings of the National Academy of Sciences of the United States of America. 74 (12): 5463–7. Bibcode:1977PNAS...74.5463S. doi:10.1073/pnas.74.12.5463. PMC 431765 . PMID 271968.
22. Maxam AM, Gilbert W (Feb 1977). "A new method for sequencing DNA". Proceedings of the National Academy of Sciences of the United States of America. 74 (2): 560–4. Bibcode:1977PNAS...74..560M. doi:10.1073/pnas.74.2.560. PMC 392330 . PMID 265521.

23. Darden L, James Tabery (2010). "Molecular Biology". In Zalta EN. The Stanford Encyclopedia of Philosophy (Fall 2010 ed.).
24. Anderson S, Bankier AT, Barrell BG, de Bruijn MH, Coulson AR, Drouin J, Eperon IC, Nierlich DP, Roe BA, Sanger F, Schreier PH, Smith AJ, Staden R, Young IG (Apr 1981). "Sequence and organization of the human mitochondrial genome". Nature. 290 (5806): 457–65. Bibcode:1981Natur.290..457A. doi:10.1038/290457a0. PMID 7219534.(subscription required)
25. Shinozaki K, Ohme M, Tanaka M, Wakasugi T, Hayashida N, Matsubayashi T, et al. (Sep 1986). "The complete nucleotide sequence of the tobacco chloroplast genome: its gene organization and expression". The EMBO Journal. 5 (9): 2043–2049. PMC 1167080 . PMID 16453699.
26. Ohyama K, Fukuzawa H, Kohchi T, Shirai H, Sano T, Sano S, et al. (1986). "Chloroplast gene organization deduced from complete sequence of liverwort Marchantia polymorpha chloroplast DNA". Nature. 322 (6079): 572–574. Bibcode:1986Natur.322..572O. doi:10.1038/322572a0.
27. Oliver SG, van der Aart QJ, Agostoni-Carbone ML, Aigle M, Alberghina L, Alexandraki D, Antoine G, Anwar R, Ballesta JP, Benit P (May 1992). "The complete DNA sequence of yeast chromosome III". Nature. 357 (6373): 38–46. Bibcode:1992Natur.357...38O. doi:10.1038/357038a0. PMID 1574125.
28. Fleischmann RD, Adams MD, White O, Clayton RA, Kirkness EF, Kerlavage AR, et al. (Jul 1995). "Whole-genome random sequencing and assembly of Haemophilus influenzae Rd". Science. 269 (5223): 496–512. Bibcode:1995Sci...269..496F. doi:10.1126/science.7542800. PMID 7542800.
29. Goffeau A, Barrell BG, Bussey H, Davis RW, Dujon B, Feldmann H, Galibert F, Hoheisel JD, Jacq C, Johnston M, Louis EJ, Mewes HW, Murakami Y, Philippsen P, Tettelin H, Oliver SG (Oct 1996). "Life with 6000 genes". Science. 274 (5287): 546, 563–7. Bibcode:1996Sci...274..546G. doi:10.1126/science.274.5287.546. PMID 8849441.(subscription required)
30. "Complete genomes: Viruses". NCBI. 2011-11-17. Retrieved 2011-11-18.
31. "Genome Project Statistics". Entrez Genome Project. 2011-10-07. Retrieved 2011-11-18.
32. Zimmer C (2009-12-29). "Scientists Start a Genomic Catalog of Earth's Abundant Microbes". The New York Times. ISSN 0362-4331. Retrieved 2012-12-21.
33. Wu D, Hugenholtz P, Mavromatis K, Pukall R, Dalin E, Ivanova NN, Kunin V, Goodwin L, Wu M, Tindall BJ, Hooper SD, Pati A, Lykidis A, Spring S, Anderson IJ, D'haeseleer P, Zemla A, Singer M, Lapidus A, Nolan M, Copeland A, Han C, Chen F, Cheng JF, Lucas S, Kerfeld C, Lang E, Gronow S, Chain P, Bruce D, Rubin EM, Kyrpides NC, Klenk HP, Eisen JA (Dec 2009). "A phylogeny-driven genomic encyclopaedia of Bacteria and Archaea". Nature. 462 (7276): 1056–60. Bibcode:2009Natur.462.1056W. doi:10.1038/nature08656. PMC 3073058 . PMID 20033048.
34. "Human gene number slashed". BBC. 2004-10-20. Retrieved 2012-12-21.
35. Yue GH, Lo LC, Zhu ZY, Lin G, Feng F (Apr 2006). "The complete nucleotide

sequence of the mitochondrial genome of Tetraodon nigroviridis". DNA Sequence. 17 (2): 115–21. doi:10.1080/10425170600700378. PMID 17076253.

36. National Human Genome Research Institute (2004-07-14). "Dog Genome Assembled: Canine Genome Now Available to Research Community Worldwide". Genome.gov. Retrieved 2012-01-20.
37. McElheny V (2010). Drawing the map of life : inside the Human Genome Project. New York NY: Basic Books. ISBN 9780465043330.
38. Abecasis GR, Auton A, Brooks LD, DePristo MA, Durbin RM, Handsaker RE, Kang HM, Marth GT, McVean GA (Nov 2012). "An integrated map of genetic variation from 1,092 human genomes". Nature. 491 (7422): 56–65. Bibcode:2012Natur.491...56T. doi:10.1038/nature11632. PMC 3498066 . PMID 23128226.
39. Nielsen R (Oct 2010). "Genomics: In search of rare human variants". Nature. 467 (7319): 1050–1. Bibcode:2010Natur.467.1050N. doi:10.1038/4671050a. PMID 20981085.
40. Barnes B, Dupré J (2008). Genomes and what to make of them. Chicago: University of Chicago Press. ISBN 978-0-226-17295-8.
41. Eisen JA (2012). "Badomics words and the power and peril of the ome-meme". GigaScience. 1 (1): 6. doi:10.1186/2047-217X-1-6. PMC 3617454 . PMID 23587201.
42. Hotz RL (2012-08-13). "Here"s an Omical Tale: Scientists Discover Spreading Suffix". Wall Street Journal. ISSN 0099-9660. Retrieved 2013-01-04.
43. Scudellari, Megan (2011-10-01). "Data Deluge". The Scientist. Retrieved 2013-01-04.
44. Chaston J, Douglas AE (Aug 2012). "Making the most of "omics" for symbiosis research". The Biological Bulletin. 223 (1): 21–9. PMC 3491573 . PMID 22983030.
45. McCutcheon JP, von Dohlen CD (Aug 2011). "An interdependent metabolic patchwork in the nested symbiosis of mealybugs". Current Biology. 21 (16): 1366–72. doi:10.1016/j.cub.2011.06.051. PMC 3169327 . PMID 21835622.
46. Monya Baker (2012-09-14). "Benchtop sequencers ship off"(Blog). Nature News Blog. Retrieved 2012-12-22.
47. Quail MA, Smith M, Coupland P, Otto TD, Harris SR, Connor TR, Bertoni A, Swerdlow HP, Gu Y (2012). "A tale of three next generation sequencing platforms: comparison of Ion Torrent, Pacific Biosciences and Illumina MiSeq sequencers". BMC Genomics. 13: 341. doi:10.1186/1471-2164-13-341. PMC 3431227 . PMID 22827831.
48. Staden R (Jun 1979). "A strategy of DNA sequencing employing computer programs". Nucleic Acids Research. 6 (7): 2601–10. doi:10.1093/nar/6.7.2601. PMC 327874 . PMID 461197.
49. Anderson S (Jul 1981). "Shotgun DNA sequencing using cloned DNase I-generated fragments". Nucleic Acids Research. 9 (13): 3015–27. doi:10.1093/nar/9.13.3015.

PMC 327328 . PMID 6269069.

50. Pop M (Jul 2009). "Genome assembly reborn: recent computational challenges". Briefings in Bioinformatics. 10 (4): 354–66. doi:10.1093/bib/bbp026. PMC 2691937 . PMID 19482960.

51. Sanger F, Coulson AR (May 1975). "A rapid method for determining sequences in DNA by primed synthesis with DNA polymerase". Journal of Molecular Biology. 94 (3): 441–8. doi:10.1016/0022-2836(75)90213-2. PMID 1100841.

52. Mavromatis K, Land ML, Brettin TS, Quest DJ, Copeland A, Clum A, et al. (2012). Liu Z, ed. "The fast changing landscape of sequencing technologies and their impact on microbial genome assemblies and annotation". PLoS ONE. 7 (12): e48837. Bibcode:2012PLoSO...748837M. doi:10.1371/journal.pone.0048837. PMC 3520994 . PMID 23251337.

53. Illumina, Inc. (2012-02-28). An Introduction to Next-Generation Sequencing Technology (PDF). San Diego, California, USA: Illumina, Inc. p. 12. Retrieved 2012-12-28.

54. Hall N (May 2007). "Advanced sequencing technologies and their wider impact in microbiology". The Journal of Experimental Biology. 210 (Pt 9): 1518–25. doi:10.1242/jeb.001370. PMID 17449817.

55. Church GM (Jan 2006). "Genomes for all". Scientific American. 294 (1): 46–54. doi:10.1038/scientificamerican0106-46. PMID 16468433.

56. ten Bosch JR, Grody WW (Nov 2008). "Keeping up with the next generation: massively parallel sequencing in clinical diagnostics". The Journal of Molecular Diagnostics. 10 (6): 484–92. doi:10.2353/jmoldx.2008.080027. PMC 2570630 . PMID 18832462.

57. Tucker T, Marra M, Friedman JM (Aug 2009). "Massively parallel sequencing: the next big thing in genetic medicine". American Journal of Human Genetics. 85 (2): 142–54. doi:10.1016/j.ajhg.2009.06.022. PMC 2725244 . PMID 19679224.

58. Kawashima EH, Farinelli L, Mayer P (2005-05-12). "Method of nucleic acid amplification". Retrieved 2012-12-22.

59. Mardis ER (2008). "Next-generation DNA sequencing methods"(PDF). Annual Review of Genomics and Human Genetics. 9: 387–402. doi:10.1146/annurev.genom.9.081307.164359. PMID 18576944.

60. Bio-IT World, Davies, K. Powering Preventative Medicine. Bio-IT World 2011

61. Chain PS, Grafham DV, Fulton RS, Fitzgerald MG, Hostetler J, Muzny D, et al. (Oct 2009). "Genomics. Genome project standards in a new era of sequencing". Science. 326 (5950): 236–7. Bibcode:2009Sci...326..236C. doi:10.1126/science.1180614. PMC 3854948 . PMID 19815760.

62. Stein L (Jul 2001). "Genome annotation: from sequence to biology". Nature Reviews Genetics. 2 (7): 493–503. doi:10.1038/35080529. PMID 11433356.

63. Brent MR (Jan 2008). "Steady progress and recent breakthroughs in the accuracy of automated genome annotation" (PDF). Nature Reviews Genetics. 9 (1): 62–73. doi:10.1038/nrg2220. PMID 18087260.

64. Flicek P, Ahmed I, Amode MR, Barrell D, Beal K, Brent S, et al. (Jan 2013). "Ensembl 2013". Nucleic Acids Research. 41 (Database issue): D48–55. doi:10.1093/nar/gks1236. PMC 3531136 . PMID 23203987.

65. Keith JM, ed. (2008). "Bioinformatics". Methods in Molecular Biology. 453. doi:10.1007/978-1-60327-429-6. ISBN 978-1-60327-428-9.

66. Marsden RL, Lewis TA, Orengo CA (2007). "Towards a comprehensive structural coverage of completed genomes: a structural genomics viewpoint". BMC Bioinformatics. 8: 86. doi:10.1186/1471-2105-8-86. PMC 1829165 . PMID 17349043.

67. Brenner SE, Levitt M (Jan 2000). "Expectations from structural genomics". Protein Science. 9 (1): 197–200. doi:10.1110/ps.9.1.197. PMC 2144435 . PMID 10739263.

68. Brenner SE (Oct 2001). "A tour of structural genomics" (PDF). Nature Reviews Genetics. 2 (10): 801–9. doi:10.1038/35093574. PMID 11584296.

69. Francis RC (2011). Epigenetics : the ultimate mystery of inheritance. New York: W.W. Norton. ISBN 9780393070057.

70. Laird PW (Mar 2010). "Principles and challenges of genomewide DNA methylation analysis". Nature Reviews Genetics. 11 (3): 191–203. doi:10.1038/nrg2732. PMID 20125086.

71. Hugenholtz P, Goebel BM, Pace NR (Sep 1998). "Impact of culture-independent studies on the emerging phylogenetic view of bacterial diversity". Journal of Bacteriology. 180 (18): 4765–74. PMC 107498 . PMID 9733676.

72. Eisen JA (Mar 2007). "Environmental shotgun sequencing: its potential and challenges for studying the hidden world of microbes". PLoS Biology. 5 (3): e82. doi:10.1371/journal.pbio.0050082. PMC 1821061 . PMID 17355177.

73. Marco D, ed. (2010). Metagenomics: Theory, Methods and Applications. Caister Academic Press. ISBN 978-1-904455-54-7.

74. Marco D, ed. (2011). Metagenomics: Current Innovations and Future Trends. Caister Academic Press. ISBN 978-1-904455-87-5.

75. Canchaya C, Proux C, Fournous G, Bruttin A, Brüssow H (Jun 2003). "Prophage genomics". Microbiology and Molecular Biology Reviews. 67 (2): 238–76, table of contents. doi:10.1128/MMBR.67.2.238-276.2003. PMC 156470 . PMID 12794192.

76. McGrath S, van Sinderen D, eds. (2007). Bacteriophage: Genetics and Molecular Biology (1st ed.). Caister Academic Press. ISBN 978-1-904455-14-1.

77. Fouts DE (November 2006). "Phage_Finder: automated identification and classification of prophage regions in complete bacterial genome sequences". Nucleic Acids Research. 34 (20): 5839–51. doi:10.1093/nar/gkl732. PMC 1635311 . PMID 17062630.

78. Herrero A, Flores E, eds. (2008). The Cyanobacteria: Molecular Biology, Genomics and Evolution (1st ed.). Caister Academic Press. ISBN 978-1-904455-15-8.

79. Hudson KL (Sep 2011). "Genomics, health care, and society". The New England

Journal of Medicine. 365 (11): 1033–41. doi:10.1056/NEJMra1010517. PMID 21916641.

80. O'Donnell CJ, Nabel EG (Dec 2011). "Genomics of cardiovascular disease". The New England Journal of Medicine. 365 (22): 2098–109. doi:10.1056/NEJMra1105239. PMID 22129254.
81. Lu YF, Goldstein DB, Angrist M, Cavalleri G (Sep 2014). "Personalized medicine and human genetic diversity". Cold Spring Harbor Perspectives in Medicine. 4 (9): a008581. doi:10.1101/cshperspect.a008581. PMID 25059740.
82. Church GM, Regis E (2012). Regenesis : how synthetic biology will reinvent nature and ourselves. New York: Basic Books. ISBN 9780465021758.
83. Baker M (May 2011). "Synthetic genomes: The next step for the synthetic genome". Nature. 473 (7347): 403, 405–8. Bibcode:2011Natur.473..403B. doi:10.1038/473403a. PMID 21593873.
84. Frankham, Richard (1 September 2010). "Challenges and opportunities of genetic approaches to biological conservation". Biological Conservation. 143 (9): 1922–1923. doi:10.1016/j.biocon.2010.05.011. Retrieved 30 May 2016.
85. Allendorf, Fred W.; Hohenlohe, Paul A.; Luikart, Gordon (October 2010). "Genomics and the future of conservation genetics". Nature Reviews Genetics. 11 (10): 697–709. doi:10.1038/nrg2844. Retrieved 30 May 2016.
86. Price DC, Chan CX, Yoon HS; et al. (2012). "Cyanophora paradoxa genome elucidates origin of photosynthesis in algae and plants". Science. 335 (6070): 843–847. Bibcode:2012Sci...335..843P. doi:10.1126/science.1213561. PMID 22344442.
87. Genome Biology | Full text | Gene functionalities and genome structure in Bathycoccus prasinos reflect cellular specializations at the base of the green lineage
88. Merchant; Prochnik, SE; Vallon, O; Harris, EH; Karpowicz, SJ; Witman, GB; Terry, A; Salamov, A; et al. (2007). "The Chlamydomonas Genome Reveals the Evolution of Key Animal and Plant Functions". Science. 318 (5848): 245–250. Bibcode:2007Sci...318..245M. doi:10.1126/science.1143609. PMC 2875087 . PMID 17932292.
89. Blanc G, Duncan G, Agarkova I, et al. (September 2010). "TheChlorella variabilis NC64A genome revals adaptation to photosymbiosis, coevolution with viruses, and cryptic sex". Plant Cell. 22 (9): 2943–2955. doi:10.1105/tpc.110.076406. PMC 2965543 . PMID 20852019.
90. Coccomyxa JGI entry
91. Smith; Lee, RW; Cushman, JC; Magnuson, JK; Tran, D; Polle, JE; et al. (2010). "The Dunaliella salina organelle genomes: large sequences, inflated with intronic and intergenic DNA". BMC Plant Biology. 10 (1): 83. doi:10.1186/1471-2229-10-83. PMC 3017802 . PMID 20459666.
92. Micromonas p.C3 JGI entry
93. ^ Jump up to:a b Worden AZ, Lee JH, Mock T, et al. (April 10, 2009).

"Green evolution and dynamic adaptations revealed by genomes of the marine picoeukaryotes Micromonas". Science. 324 (5924): 268–272. Bibcode:2009Sci...324..268W. doi:10.1126/science.1167222. PMID 19359590.

94. Micromonas p.N3 JGI entry
95. Palenik, B; Grimwood, J; Aerts, A; Rouzé, P; Salamov, A; Putnam, N; Dupont, C; Jorgensen, R; et al. (2007). "The tiny eukaryote Ostreococcus provides genomic insights into the paradox of plankton speciation". Proceedings of the National Academy of Sciences of the United States of America. 104 (18): 7705–10. Bibcode:2007PNAS..104.7705P. doi:10.1073/pnas.0611046104. PMC 1863510 . PMID 17460045.
96. Derelle E, Ferraz C, Rombauts S, et al. (August 2006). "Genome analysis of the smallest free-living eukaryote Ostreococcus tauri unveils many unique features". PNAS. 103 (31): 11647–52. Bibcode:2006PNAS..10311647D. doi:10.1073/pnas.0604795103. PMC 1544224 . PMID 16868079.
97. Info - Ostreococcus RCC809
98. Prochnik SE, Umen J, Nedelcu AM; Umen; Nedelcu; Hallmann; Miller; Nishii; Ferris; Kuo; et al. (July 9, 2010). "Genomic analysis of organismal complexity in the multicellular green alga Volvox carteri". Science. 329 (5988): 223–226. Bibcode:2010Sci...329..223P. doi:10.1126/science.1188800. PMC 2993248 . PMID 20616280.
99. Collén, J.; Porcel, B.; Carré, W.; Ball, S. G.; Chaparro, C.; Tonon, T.; Boyen, C. (2013). "Genome structure and metabolic features in the red seaweed Chondrus crispus shed light on evolution of the Archaeplastida". Proceedings of the National Academy of Sciences. 110: 5247–5252. doi:10.1073/pnas.1221259110.
100. Matsuzaki M, Misumi O, Shin-I T, et al. (April 2004). "Genome sequence of the ultrasmall unicellular red alga Cyanidioschyzon merolae 10D". Nature. 428 (6983): 653–7. Bibcode:2004Natur.428..653M. doi:10.1038/nature02398. PMID 15071595.
101. Nozaki; et al. (2007). "A 100%-complete sequence reveals unusually simple genomic features in the hot-spring red alga Cyanidioschyzon merolae". BMC Biol. 5: 28. doi:10.1186/1741-7007-5-28.
102. Galdieria sulphuraria Genome Project at MSU
103. Barbier, G; Oesterhelt, C; Larson, MD; et al. (February 2005). "Comparative genomics of two closely related unicellular thermo-acidophilic red algae, Galdieria sulphuraria and Cyanidioschyzon merolae, reveals the molecular basis of the metabolic flexibility of Galdieria sulphuraria and significant differences in carbohydrate metabolism of both algae". Plant Physiol. 137: 460–74. doi:10.1104/pp.104.051169. PMC 1065348 . PMID 15710685.
104. Schönknecht; et al. (Mar 2013). "(March 8, 2013) Gene Transfer from Bacteria and Archaea Facilitated Evolution of an Extremophilic Eukaryote". Science. 339: 1207–1210. doi:10.1126/science.1231707. PMID 23471408.
105. Bhattacharya; et al. (2013). "Genome of the red alga Porphyridium purpureum". Nature Communications. 4: 1941. doi:10.1038/ncomms2931.

106. Nakamura; et al. "(March 11, 2013) The First Symbiont-Free Genome Sequence of Marine Red Alga, Susabi-nori (Pyropia yezoensis)". PLoS ONE. 8 (3): e57122. doi:10.1371/journal.pone.0057122.

107. Cock JM, Sterck L, Rouzé P; Sterck; Rouzé; Scornet; Allen; Amoutzias; Anthouard; Artiguenave; et al. (June 3, 2010). "The Ectocarpus genome and the independent evolution of multicellularity in brown algae". Nature. 465 (7298): 617–621. Bibcode:2010Natur.465..617C. doi:10.1038/nature09016. PMID 20520714.

108. Rensing SA, Lang D, Zimmer AD, et al. (January 2008). "The Physcomitrella genome reveals evolutionary insights into the conquest of land by plants". Science. 319 (5859): 64–9. Bibcode:2008Sci...319...64R. doi:10.1126/science.1150646. PMID 18079367.

109. Banks JA, Nishiyama T, Hasebe M, et al. (20 May 2011). "TheSelaginella genome identifies genetic changes associated with the evolution of vascular plants". Science. 332 (6032): 960–963. Bibcode:2011Sci...332..960B. doi:10.1126/science.1203810. PMC 3166216 . PMID 21551031.

110. JGI project page

111. Amborella Genome Project (20 Dec 2013). "The Amborella genome and the evolution of flowering plants". Science. 342 (6165): 1241089. doi:10.1126/science.1241089. PMID 24357323.

112. amborella.org

113. Phytozome v9.1: Aquilegia caerulea

114. Ray Ming; Robert VanBuren; Yanling Liu; Mei Yang; Yuepeng Han; et al. (10 May 2013). "Genome of the long-living sacred lotus (Nelumbo nucifera Gaertn.)". Genome Biology. 14 (5): R41. doi:10.1186/gb-2013-14-5-r41.

115. Juliane C. Dohm; André E. Minoche; Daniela Holtgräwe; Salvador Capella-Gutiérrez; Falk Zakrzewski; Hakim Tafer; Oliver Rupp; Thomas Rosleff Sörensen; Ralf Stracke; Richard Reinhardt; Alexander Goesmann; Thomas Kraft; Britta Schulz; Peter F. Stadler; Thomas Schmidt; Toni Gabaldón; Hans Lehrach; Bernd Weisshaar; Heinz Himmelbauer (December 2013). "The genome of the recently domesticated crop plant sugar beet (Beta vulgaris)". Nature. 505 (7484): 546–549. doi:10.1038/nature12817.

116. Wang N, Thomson M, Bodles W, et al. (2013). "Genome sequence of dwarf birch (Betula nana) and cross-species RAD markers". Molecular Ecology. 22 (11): 3098–3111. doi:10.1111/mec.12131. PMID 23167599.

117. An atlas of over 90,000 conserved noncoding sequences provides insight into crucifer regulatory regions : Nature Genetics : Nature Publishing Group

118. Hu T, Pattyn P, Bakker EG, et al. (April 2011). "The Arabidopsis lyrata genome sequence and the basis of rapid genome size change". Nature Genetics. 43 (5): 476–81. doi:10.1038/ng.807.

119. The Arabidopsis Genome Initiative (December 2000). "Analysis of the genome sequence of the flowering plant Arabidopsis thaliana". Nature. 408 (6814): 796–815. doi:10.1038/35048692. PMID 11130711.

120. Wang X, Wang H, Wang J, et al. (2011). "The genome of the mesopolypoid crop species Brassica rapa". Nature Genetics. 43: 1035–1039. doi:10.1038/ng.919. PMID 21873998.

121. Phytozome v9.1: Capsella rubella

122. Slotte T, et al. (2013). "The Capsella rubella genome and the genomic consequences of rapid mating system evolution.". Nature Genetics. 45 (7): 831–835. doi:10.1038/ng.2669. PMID 23749190.

123. Yang, R; Jarvis, DE; Chen, H; Beilstein, MA; Grimwood, J; Jenkins, J; Shu, S; Prochnik, S; Xin, M; et al. "The Reference Genome of the Halophytic Plant Eutrema salsugineum". Front Plant Sci. 4: 46. doi:10.3389/fpls.2013.00046. PMC 3604812 . PMID 23518688.

124. Dassanayake M, Oh DH, Haas JS, et al. (2011). "The genome of the extremophile crucifer Thellungiella parvula". Nature Genetics. 43 (9): 913–918. doi:10.1038/ng.889.

125. van Bakel H, Stout JM, Cote AT, et al. (2011). "The draft genome and transcriptome of Cannabis sativa". Genome Biology. 12 (10): R102. doi:10.1186/gb-2011-12-10-r102. PMC 3359589 . PMID 22014239.

126. "The draft genome of the transgenic tropical fruit tree papaya (Carica papaya Linnaeus)". Nature. 452: 991–6. doi:10.1038/nature06856. PMC 2836516 . PMID 18432245.

127. Kalanchoe

128. The draft genome of watermelon (Citrullus lanatus) and resequencing of 20 diverse accessions : Nature Genetics : Nature Publishing Group

129. The genome of melon (Cucumis melo L.)

130. Huang S, Li R, Zhangl Z; et al. (December 2009). "The genome of the cucumber, Cucumis sativus L". Nature Genetics. 41 (12): 1275–. doi:10.1038/ng.475. PMID 19881527.

131. BMC Genomics | Full text | Draft genome sequence of the rubber tree Hevea brasiliensis

132. Sato S, Hirakawa H, Isobe S, et al. (February 2011). "Sequence analysis of the genome of an oil-bearing tree, Jatropha curcas L.". DNA Research. 18 (1): 65–76. doi:10.1093/dnares/dsq030. PMC 3041505 . PMID 21149391.

133. Prochnik et al. (2012), J. Tropical Plant Biology

134. Chan AP, Crabtree J, Zhao Q, et al. (2010). "Draft genome sequence of the oilseed species Ricinus communis". Nature Biotechnology. 28 (9): "951–956". doi:10.1038/nbt.1674.

135. Nagendra K. Singh; Deepak K. Gupta; Pawan K. Jayaswal; Ajay K. Mahato; Sutapa Dutta; Sangeeta Singh; Shefali Bhutani; Vivek Dogra; Bikram P. Singh; Giriraj Kumawat; et al. (2012). "The first draft of the pigeonpea genome sequence". Journal of Plant Biochemistry and Biotechnology. 21 (1): 98–112. doi:10.1007/s13562-011-0088-8. PMC 3886394 . PMID 24431589.

136. Varshney RK, Chen W, Li Y, et al. (2012). "Draft genome sequence of pigeonpea

(Cajanus cajan), an orphan legume crop of resource-poor farmers". Nature Biotechnology. 30 (1): 83–89. doi:10.1038/nbt.2022. PMID 22057054.

137. Bertioli, David John; Cannon, Steven B.; Froenicke, Lutz; Huang, Guodong; Farmer, Andrew D.; Cannon, Ethalinda K. S.; Liu, Xin; Gao, Dongying; Clevenger, Josh (2016-04-01). "The genome sequences of Arachis duranensis and Arachis ipaensis, the diploid ancestors of cultivated peanut". Nature Genetics. 48 (4): 438–446. doi:10.1038/ng.3517. ISSN 1061-4036.
138. Draft genome sequence of chickpea (Cicer arietinum) provides a resource for trait improvement : Nature Biotechnology : Nature Publishing Group
139. A draft genome sequence of the pulse crop chickpea (Cicer arietinum L.) - Jain - 2013 - The Plant Journal - Wiley Online Library
140. Huang S, Li R, Zhang1 Z; et al. (10 January 2010). "Genome sequence of the palaeopolyploid soybean". Nature. 463 (12): 178–183. Bibcode:2010Natur.463..178S. doi:10.1038/nature08670. PMID 20075913.
141. Sato S, Nakamura Y, Kaneko T, et al. (2008). "Genome structure of the legume, Lotus japonicus". DNA Research. 15 (4): 227–239. doi:10.1093/dnares/dsn008. PMC 2575887 . PMID 18511435.
142. Young ND, Debelle F, Oldroyd GE, et al. (2011). "The Medicago genome provides insight into the evolution of rizobial symbioses". Nature. 480 (7378). Bibcode:2011Natur.480..520Y. doi:10.1038/nature10625. PMC 3272368 . PMID 22089132.
143. Phytozome v9.1: Phaseolus vulgaris v1.0
144. "The genome of flax (Linum usitatissimum) assembled de novo from short shotgun sequence reads". The Plant Journal. 72: 461–473. doi:10.1111/j.1365-313X.2012.05093.x.
145. Phytozome v9.1: Gossypium raimondii v2.1
146. Argout, Xavier; Salse, Jerome; Aury, Jean-Marc; Guiltinan, Mark J; Droc, Gaetan; Gouzy, Jerome; Allegre, Mathilde; Chaparro, Cristian; et al. (2010). "The genome of Theobroma cacao". Nature Genetics. 43 (2): 101–8. doi:10.1038/ng.736. PMID 21186351.
147. Pennisi, E. (2010). "Genomics Researchers Upset by Rivals' Publicity". Science. 329 (5999): 1585. Bibcode:2010Sci...329.1585P. doi:10.1126/science.329.5999.1585. PMID 20929817.
148. Genome Biology | Abstract | The genome sequence of the most widely cultivated cacao type and its use to identify candidate genes regulating pod color
149. BMC Genomics | Full text | A draft of the genome and four transcriptomes of a medicinal and pesticidal angiosperm Azadirachta indica
150. http://www.currentscience.ac.in/Volumes/101/12/1553.pdf
151. Myburg, Alexander A.; Grattapaglia, Dario; Tuskan, Gerald A.; Hellsten, Uffe; Hayes, Richard D.; Grimwood, Jane; Jenkins, Jerry; Lindquist, Erika; Tice, Hope. "The genome of Eucalyptus grandis". Nature. doi:10.1038/nature13308.
152. Shulaev V, Sargent DJ, Crowhurst RN, et al. (2011). "The genome of woodland

strawberry (Fragaria vesca)". Nature Genetics. 43 (2): 109–116. doi:10.1038/ng.740. PMC 3326587 . PMID 21186353.

153. Velasco, R.; Zharkikh, A.; Affourtit, J.; Dhingra, A.; Cestaro, A.; Kalyanaraman, A.; Fontana, P.; Bhatnagar, S.; Troggio, M.; Pruss, D.; Salvi, S.; Pindo, M.; Baldi, P.; Castelletti, S.; Cavaiuolo, M.; Coppola, G.; Costa, F.; Cova, V.; Dal Ri, A.; Goremykin, V.; Komjanc, M.; Longhi, S.; Magnago, P.; Malacarne, G.; Malnoy, M.; Micheletti, D.; Moretto, M.; Perazzolli, M.; Si-Ammour, A.; Vezzulli, S. (2010). "The genome of the domesticated apple (Malus x domestica Borkh.)". Nature Genetics. 42 (10): 833–839. doi:10.1038/ng.654. PMID 20802477.
154. Gramene News " Blog Archive " Four Rosaceae Genomes Released
155. The genome of Prunus mume : Nature Communications : Nature Publishing Group
156. The International Peach Genome Initiative; et al. (2013). "The high-quality draft genome of peach (Prunus persica) identifies unique patterns of genetic diversity, domestication and genome evolution". Nature Genetics. 45 (5): 487–494. doi:10.1038/ng.2586. PMID 23525075.
157. The genome of the pear (Pyrus bretschneideri Rehd.)
158. Phytozome v9.1: Citrus clementina
159. Xu Q, et al. (2013). "The draft genome of sweet orange (Citrus sinensis)". Nature Genetics. 45 (1): 59–66. doi:10.1038/ng.2472. PMID 23179022.
160. Tuskan GA, Difazio S, Jansson S, et al. (September 2006). "The genome of black cottonwood, Populus trichocarpa (Torr. & Gray)". Science. 313 (5793): 1596–604. Bibcode:2006Sci...313.1596T. doi:10.1126/science.1128691. PMID 16973872.
161. Jaillon O, Aury JM, Noel B, et al. (September 2007). "The grapevine genome sequence suggests ancestral hexaploidization in major angiosperm phyla". Nature. 449 (7161): 463–7. Bibcode:2007Natur.449..463J. doi:10.1038/nature06148. PMID 17721507.
162. Phytozome v9.1: Mimulus guttatus
163. Sol Genomics Network
164. The Tomato Genome Consortium (31 May 2012). "The tomato genome sequence provides insights into fleshy fruit evolution". Nature. 485 (7400): 635–641. Bibcode:2012Natur.485..635T. doi:10.1038/nature11119. PMC 3378239 . PMID 22660326.
165. Xu, X.; Xu, S.; Pan, S.; Cheng, B.; Zhang, D.; Mu, P.; Ni, G.; Zhang, S.; Yang, R.; Li, J.; Wang, G.; Orjeda, F.; Guzman, M.; Torres, R.; Lozano, O.; Ponce, D.; Martinez, G. N.; De La Cruz, S. K.; Chakrabarti, V. U.; Patil, K. G.; Skryabin, B. B.; Kuznetsov, N. V.; Ravin, T. V.; Kolganova, A. V.; Beletsky, A. V.; Mardanov, A.; Di Genova, D. M.; Bolser, D. M. A.; Martin, G.; Li, Y. (2011). "Genome sequence and analysis of the tuber crop potato". Nature. 475 (7355): 189–195. doi:10.1038/nature10158. PMID 21743474.
166. Aversano R, Contaldi F, Ercolano MR, Grosso V, Iorizzo M, Tatino F, Xumerle L, Dal Molin A, Avanzato C, Ferrarini A, Delledonne M, Sanseverino W, Aiese

Cigliano R, Capella-Gutierrez S, Gabaldón T, Frusciante L, Bradeen JM, Carputo D. (14 Apr 2015). "The Solanum commersonii genome sequence provides insights into adaptation to stress conditions and genome evolution of wild potato relatives". The Plant Cell. 27 (4): 954–968. doi:10.1105/tpc.114.135954. PMID 25873387.

167. Bombarely, A; Rosli, HG; Vrebalov, J; Moffett, P; Mueller, LA; Martin, GB. "A draft genome sequence of Nicotiana benthamiana to enhance molecular plant-microbe biology research". Mol Plant Microbe Interact. 25: 1523–30. doi:10.1094/MPMI-06-12-0148-TA. PMID 22876960.
168. Genome Biology | Abstract | Reference genomes and transcriptomes of Nicotiana sylvestris and Nicotiana tomentosiformis
169. Kim; et al. (2014). "Genome sequence of the hot pepper provides insights into the evolution of pungency in Capsicum species". Nature Genetics. 46 (3): 270–278. doi:10.1038/ng.2877.
170. Qin; et al. (2014). "Whole-genome sequencing of cultivated and wild peppers provides insights into Capsicum domestication and specialization.". Proc Natl Acad Sci U S A. 111: 5135–5140. doi:10.1073/pnas.1400975111.
171. Bennetzen, JL; Schmutz, J; Wang, H; Percifield, R; Hawkins, J; Pontaroli, AC; Estep, M; Feng, L; Vaughn, JN; et al. "Reference genome sequence of the model plant Setaria". Nat Biotechnol. 30: 555–61. doi:10.1038/nbt.2196. PMID 22580951.
172. Jia J, Zhou S, Kong X, et al. (2013). "Aegilops tauschii draft genome sequence reveals a gene repertoire for wheat adaptation". Nature. 496 (7443): 91–95. Bibcode:2013Natur.496...91.. doi:10.1038/nature12028.
173. The International Brachypodium Initiative (December 2009). "Genome sequencing and analysis of the model grass Brachypodium distachyon". Nature. 463 (7282): 763–8. Bibcode:2010Natur.463..763T. doi:10.1038/nature08747. PMID 20148030.
174. The International Barley Genome Sequencing Consortium. "A physical, genetic and functional sequence assembly of the barley genome". Nature. 491 (7426): 711–6. Bibcode:2012Natur.491..711T. doi:10.1038/nature11543. PMID 23075845.
175. Whole-genome sequencing of Oryza brachyantha reveals mechanisms underlying Oryza genome evolution : Nature Communications : Nature Publishing Group
176. Hurwitz, BL; Kudrna, D; Yu, Y; Sebastian, A; Zuccolo, A; Jackson, SA; Ware, D; Wing, RA; et al. (2010). "Rice structural variation: A comparative analysis of structural variation between rice and three of its closest relatives in the genus Oryza". The Plant journal : for cell and molecular biology. 63 (6): 990–1003. doi:10.1111/j.1365-313X.2010.04293.x. PMID 20626650.
177. Huang, X.; Kurata, N.; Wei, X.; Wang, Z. X.; Wang, A.; Zhao, Q.; Zhao, Y.; Liu, K.; Lu, H.; Li, W.; Guo, Y.; Lu, Y.; Zhou, C.; Fan, D.; Weng, Q.; Zhu, C.; Huang, T.; Zhang, L.; Wang, Y.; Feng, L.; Furuumi, H.; Kubo, T.; Miyabayashi, T.; Yuan, X.; Xu, Q.; Dong, G.; Zhan, Q.; Li, C.; Fujiyama, A.; et al. (2012). "A map of rice genome variation reveals the origin of cultivated rice". Nature. 490 (7421): 497–501. doi:10.1038/nature11532. PMID 23034647.

178. Yu J, Hu S, Wang J, et al. (April 2002). "A draft sequence of the rice genome (Oryza sativa L. ssp. indica)". Science. 296 (5565): 79–92. Bibcode:2002Sci...296...79Y. doi:10.1126/science.1068037. PMID 11935017.

179. Goff SA, Ricke D, Lan TH, et al. (April 2002). "A draft sequence of the rice genome (Oryza sativa L. ssp. japonica)". Science. 296 (5565): 92–100. Bibcode:2002Sci...296...92G. doi:10.1126/science.1068275. PMID 11935018.

180. Phytozome v9.1: Panicum virgatum

181. The draft genome of the fast-growing non-timber forest species moso bamboo (Phyllostachys heterocycla) : Nature Genetics : Nature Publishing Group

182. Paterson, A.; Bowers, J.; Bruggmann, R.; Dubchak, I.; Grimwood, J.; Gundlach, H.; Haberer, G.; Hellsten, U.; Mitros, T.; Poliakov, A.; Schmutz, J.; Spannagl, M.; Tang, H.; Wang, X.; Wicker, T.; Bharti, A. K.; Chapman, J.; Feltus, F. A.; Gowik, U.; Grigoriev, I. V.; Lyons, E.; Maher, C. A.; Martis, M.; Narechania, A.; Otillar, R. P.; Penning, B. W.; Salamov, A. A.; Wang, Y.; Zhang, L.; et al. (2009). "The Sorghum bicolor genome and the diversification of grasses" (PDF). Nature. 457 (7229): 551–556. Bibcode:2009Natur.457..551P. doi:10.1038/nature07723. PMID 19189423.

183. Brenchley R, Spannagl M, Pfeifer M, et al. (2012). "Analysis of the bread wheat genome using whole-genome shotgun sequencing". Nature. 491 (7426): 705–710. Bibcode:2012Natur.491..705B. doi:10.1038/nature11650. PMC 3510651 . PMID 23192148.

184. Ling H-Q, Zhou S, Liu D et al. (2013). "Draft genome of the wheat A-genome progenitor Triticum urartu". Nature. 496 (7443): 87–90. Bibcode:2013Natur.496...87L. doi:10.1038/nature11997.

185. MaizeSequence 5b.60: Home

186. Schnable P, Ware D, Fulton RS, et al. (22 November 2009). "The B73 Maize Genome: Complexity, Diversity, and Dynamics". Science. 326 (5956): 1112–1115. Bibcode:2009Sci...326.1112S. doi:10.1126/science.1178534. PMID 19965430.

187. D'Hont; et al. (2012). "The banana (Musa acuminata) genome and the evolution of monocotyledonous plants". Nature. 488 (7410): 213–217. Bibcode:2012Natur.488..213D. doi:10.1038/nature11241. PMID 22801500.

188. Davey; et al. (2013). "A draft Musa balbisiana genome sequence for molecular genetics in polyploid, inter- and intra-specific Musa hybrids.". BMC Genomics. 14: 683. doi:10.1186/1471-2164-14-683.

189. Al-Dous; et al. (2011). "De novo genome sequencing and comparative genomics of date palm (Phoenix dactylifera)". Nature Biotechnology. 29 (6): 521–527. doi:10.1038/nbt.1860.

190. Singh; et al. (2013). "Oil palm genome sequence reveals divergence of interfertile species in Old and New worlds.". Nature. 500 (7462): 335–339. doi:10.1038/nature12309.

191. Wang; et al. (2014). "The Spirodela polyrhiza genome reveals insights into its neotenous reduction fast growth and aquatic lifestyle.". Nature Commun. 5:

3311. doi:10.1038/ncomms4311. PMC 3948053 . PMID 24548928.

192. Nystedt B, Street NR, Wetterbom A, Zuccolo A, Lin YC, Scofield DG, Vezzi F, Delhomme N, Giacomello S, Alexeyenko A, Vicedomini R, Sahlin K, Sherwood E, Elfstrand M, Gramzow L, Holmberg K, Hällman J, Keech O, Klasson L, Koriabine M, Kucukoglu M, Käller M, Luthman J, Lysholm F, Niittylä T, Olson A, Rilakovic N, Ritland C, Rosselló JA, Sena J, Svensson T, Talavera-López C, Theißen G, Tuominen H, Vanneste K, Wu ZQ, Zhang B, Zerbe P, Arvestad L, Bhalerao R, Bohlmann J, Bousquet J, Garcia Gil R, Hvidsten TR, de Jong P, MacKay J, Morgante M, Ritland K, Sundberg B, Thompson SL, Van de Peer Y, Andersson B, Nilsson O, Ingvarsson PK, Lundeberg J, Jansson S (2013). "The Norway spruce genome sequence and conifer genome evolution". Nature. 497 (7451): 579–584. doi:10.1038/nature12211. PMID 23698360.

193. Birol I, Raymond A, Jackman SD, Pleasance S, Coope R, Taylor GA, Yuen MM, Keeling CI, Brand D, Vandervalk BP, Kirk H, Pandoh P, Moore RA, Zhao Y, Mungall AJ, Jaquish B, Yanchuk A, Ritland C, Boyle B, Bousquet J, Ritland K, Mackay J, Bohlmann J, Jones SJ (2013). "Assembling the 20 Gb white spruce (Picea glauca) genome from whole-genome shotgun sequencing data". Bioinformatics. 29 (12): 1492–1497. doi:10.1093/bioinformatics/btt178. PMC 3673215 . PMID 23698863.

194. Zimin A, Stevens KA, Crepeau MW, Holtz-Morris A, Koriabine M, Marçais G, Puiu D, Roberts M, Wegrzyn JL, de Jong PJ, Neale DB, Salzberg SL, Yorke JA, Langley CH (2014). "Sequencing and Assembly of the 22-Gb Loblolly Pine Genome". Genetics. 196 (3): 875–890. doi:10.1534/genetics.113.159715. PMC 3948813 . PMID 24653210.

195. Wegrzyn JL, Liechty JD, Stevens KA, Wu LS, Loopstra CA, Vasquez-Gross HA, Dougherty WM, Lin BY, Zieve JJ, Martínez-García PJ, Holt C, Yandell M, Zimin AV, Yorke JA, Crepeau MW, Puiu D, Salzberg SL, Dejong PJ, Mockaitis K, Main D, Langley CH, Neale DB (2014). "Unique Features of the Loblolly Pine (Pinus taeda L.) Megagenome Revealed Through Sequence Annotation". Genetics. 196 (3): 891–909. doi:10.1534/genetics.113.159996. PMC 3948814 . PMID 24653211.

196. Neale DB, Wegrzyn JL, Stevens KA, Zimin AV, Puiu D, Crepeau MW, Cardeno C, Koriabine M, Holtz-Morris AE, Liechty JD, Martínez-García PJ, Vasquez-Gross HA, Lin BY, Zieve JJ, Dougherty WM, Fuentes-Soriano S, Wu LS, Gilbert D, Marçais G, Roberts M, Holt C, Yandell M, Davis JM, Smith KE, Dean JF, Lorenz WW, Whetten RW, Sederoff R, Wheeler N, McGuire PE, Main D, Loopstra CA, Mockaitis K, deJong PJ, Yorke JA, Salzberg SL, Langley CH (2014). "Decoding the massive genome of loblolly pine using haploid DNA and novel assembly strategies". Genome Biology. 15 (3): R59. doi:10.1186/gb-2014-15-3-r59. PMC 4053751 . PMID 24647006.

197. http://www.research-in-germany.de/coremedia/generator/dachportal/en/07__News_20and_20Events/VDITZ_20-_20News_26Events/Archiv/2009-10-25_2C_20Full_20oilseed_20rape_20genome_20deciphered,sourcePageId=34814.html

198. "First Draft Of Oil Palm Genome Completed". Energy-daily.com. Retrieved 2010-08-27.

199. "Jute genome decoded : Golden fibre to become healthy, high yielding, weather-tolerant; Hawaii-based Bangladeshi scientist leads team to landmark discovery". The Daily Star.

200. "Jute genome sequence decoded by Bangladeshi scientists". Hindusthan Times.

201. "স্‌বপ্‌নযাত্‌রা (Chasing the dream)". Jute Genome Project.

202. Welcome to the British Ash Tree Genome Project | The British Ash Tree Genome Project -

203. The School of Biological & Chemical Sciences

204. BBC News - Ash genome reveals fungus resistance

CHAPTER 4

Doubled Haploidy

Chapter Contents

- **Introduction**
- **Production of Doubled Haploids**
- **Genetics of DH Population**
- **Applications of DHS Plant Breeding**
- **Advantages of DHS**
- **Disadvantages of DHS**

For thousands of years, farmers have worked to improve their crop varieties, slowly breeding new traits over many generations. Today, advanced plant breeding technologies enable far more significant improvements in far less time. Among the most promising emerging technologies is the production of doubled haploids – genetically pure plant lines that offer a quick route to new gene combinations for higher yield, improved quality, disease resistance, or other crop improvements.

INTRODUCTION

A doubled haploid (DH) is a genotype formed when haploid cells undergo chromosome doubling. Artificial production of doubled haploids is important in plant breeding.

Haploid cells are produced from pollen or egg cells or from other cells of the gametophyte, then by induced or spontaneous chromosome doubling, a doubled haploid cell is produced, which can be grown into a doubled haploid plant. If the original plant was diploid, the haploid cells are monoploid, and the term **doubled monoploid** may be used for the doubled haploids. Haploid organisms derived from tetraploids or hexaploids are sometimes called dihaploids (and the doubled dihaploids are, respectively, tetraploid or hexaploid).

Conventional inbreeding procedures take six generations to achieve approximately complete homozygosity, whereas doubled haploidy achieves it in one generation. Dihaploid plants derived from tetraploid crop plants may be important for breeding programs that involve diploid wild relatives of the crops.

The first report of the haploid plant was published by Blakeslee *et al.* (1922) in *Datura stramonium*. Subsequently, haploids were reported in many other species. Guha and Maheshwari (1964) developed an anther culture technique for the production of haploids in the laboratory. Haploid production by wide crossing was reported in barley (Kasha and Kao, 1970) and tobacco (Burk *et al.*, 1979). Tobacco, rapeseed, and barley are the most responsive species for doubled haploid production. Doubled haploid methodologies have now been applied to over 250 species.

PRODUCTION OF DOUBLED HAPLOIDS

Doubled haploids can be produced *in vivo* or *in vitro*. Haploid embryos are produced *in vivo* by parthenogenesis, pseudogamy, or chromosome elimination after wide crossing. The haploid embryo is rescued, cultured, and chromosome-doubling produces doubled haploids. The *in vitro* methods include gynogenesis (ovary and flower culture) and androgenesis (anther and microspore culture). Androgenesis is the preferred method. Another method of producing the haploids is wide crossing. In barley, haploids can be produced by wide crossing with the related species *Hordeum bulbosum*; fertilization is affected, but during the early stages of seed development the *H. bulbosum* chromosomes are eliminated leaving a haploid embryo. In tobacco (*Nicotiana tabacum*), wide crossing with *Nicotiana africana* is widely used. When *N. africana* is used to pollinate *N. tabacum*, 0.25 to 1.42 percent of the progeny survive and can readily be identified as either F1 hybrids or maternal haploids. Although these percentages appear small, the vast yield of tiny seeds and the early death of most seedlings provide significant numbers of viable hybrids and haploids in relatively small soil containers. This method of interspecific pollination serves as a practical way of producing seed-derived haploids of *N. tabacum*, either as an alternative method or complementary method to anther culture.

GENETICS OF DH POPULATION

In DH method only two types of genotypes occur for a pair of alleles, A and a, with the frequency of ½ AA and ½ aa, while in diploid method three genotypes occur with the frequency of ¼ AA, ½ Aa, ¼ aa. Thus, if AA is desirable genotype, the probability of obtaining this genotype is higher in haploid method than in diploid method. If n loci are segregating, the probability of getting the desirable genotype is (1/2)n by the haploid method and (1/4)n by the diploid method. Hence the efficiency of the haploid method is high when the number of genes concerned is large.

Studies were conducted comparing DH method and other conventional breeding methods and it was concluded that adoption of doubled haploidy does not lead to any bias of genotypes in populations, and random DHs were even found to be compatible to selected line produced by conventional pedigree method.

APPLICATIONS OF DHS PLANT BREEDING

Mapping Quantitative Trait Loci

Most of the economic traits are controlled by genes with small but cumulative effects. Although the potential of DH populations in quantitative genetics has been understood for some time, it was the advent of molecular marker maps that provided the impetus for their use in identifying loci controlling quantitative traits. As the quantitative trait loci (QTL) effects are small and highly influenced by environmental factors, accurate phenotyping with replicated trials is needed. This is possible with doubled haploidy organisms because of their true breeding nature and because they can conveniently be produced in large numbers. Using DH populations, 130 quantitative traits have been mapped in nine crop species. In total, 56 DH populations were used for QTL detection.

Backcross Breeding

In backcross conversion, genes are introgressed from a donor cultivar or related species into a recipient elite line through repeated backcrossing. A problem in this procedure is being able to identify the lines carrying the trait of interest at each generation. The problem is particularly acute if the trait of interest is recessive, as it will be present only in a heterozygous condition after each backcross. The development of molecular markers provides an easier method of selection based on the genotype (marker) rather than the phenotype. Combined with doubled haploidy it becomes more effective. In marker assisted backcross conversion, a recipient parent is crossed with a donor line and the hybrid (F1) backcrossed to the recipient. The resulting generation (BC1) is backcrossed and the process repeated until the desired genotypes are produced. The combination of doubled haploidy and molecular marker provides the short cut. In the back cross generation one itself a genotype with the character of interest can be selected and converted into homozygous doubled haploid genotype. Chen *et al.* (1994) used marker assisted backcross conversion with doubled haploidy of BC1 individuals to select stripe rust resistant lines in barley.

Bulked Segregant Analysis (BSA)

In bulked segregant analysis, a population is screened for a trait of interest and the genotypes at the two extreme ends form two bulks. Then the two bulks are tested for the presence or absence of molecular markers. Since the bulks are supposed to contrast in the alleles that contribute positive and negative effects, any marker polymorphism between the two bulks indicates the linkage between the marker and trait of interest. BSA is dependent on accurate phenotyping and the DH population has particular advantage in that they are true breeding and can be tested repeatedly. DH populations are commonly used in bulked segregant analysis, which is a popular method in marker assisted breeding. This method has been applied mostly to rapeseed and barley.

Genetic Maps

Genetic maps are very important to understand the structure and organization of genomes from which evolution patterns and syntenic relationships between species can be deduced. Genetic maps also provide a framework for the mapping of genes of interest and estimating the magnitude of their effects and aid our understanding of genotype/phenotype associations. DH populations have become standard resources in genetic mapping for species in which DHs are readily available. Doubled haploid populations are ideal for genetic mapping. It is possible to produce a genetic map within two years of the initial cross regardless of the species. Map construction is relatively easy using a DH population derived from a hybrid of two homozygous parents as the expected segregation ratio is simple, *i.e.* 1:1. DH populations have now been used to produce genetic maps of barley, rapeseed, rice, wheat, and pepper. DH populations played a major role in facilitating the generation of the molecular marker maps in eight crop species.

Genetic Studies

Genetic ratios and mutation rates can be read directly from haploid populations. A small doubled haploid (DH) population was used to demonstrate that a dwarfing gene in barley is located chromosome 5H. In another study the segregation of a range of markers has been analyzed in barley.

Genomics

Although QTL analysis has generated a vast amount of information on gene locations and the magnitude of effects on many traits, the identification of the genes involved has remained elusive. This is due to poor resolution of QTL analysis. The solution for this problem would be production of recombinant chromosome substitution line, or stepped aligned recombinant inbred lines. Here, backcrossing is carried out until a desired level of recombination has occurred and genetic markers are used to detect desired recombinant chromosome substitution lines in the target region, which can be fixed by doubled haploidy. In rice, molecular markers have been found to be linked with major genes and QTLs for resistance to rice blast, bacterial blight, and sheath blight in a map produced from DH population.

Elite Crossing

Traditional breeding methods are slow and take 10–15 years for cultivar development. Another disadvantage is inefficiency of selection in early generations because of heterozygosity. These two disadvantages can be over come by DHs, and more elite crosses can be evaluated and selected within less time.

Cultivar Development

Uniformity is a general requirement of cultivated line in most species, which can be easily obtained through DH production. There are various ways in which DHs can be used in cultivar production. The DH lines themselves can be released as cultivars, they may be used as parents in hybrid cultivar production or more indirectly in the creation of breeders lines and in germplasm conservation. Barley has over 100 direct DH cultivars. According to published information there are currently around 300 DH derived cultivars in 12 species worldwide.

The relevance of DHs to plant breeding has increased markedly in recent years owing to the development of protocols for 25 species. Doubled haploidy already plays an important role in hybrid cultivar production of vegetables, and the potential for ornamental production is being vigorously examined. DHs are also being developed in the medicinal herb *Valeriana officinalis* to select lines with high pharmacological activity. Another interesting development is that fertile homozygous DH lines can be produced in species that have self-incompatibility systems.

ADVANTAGES OF DHS

The ability to produce homozygous lines after a single round recombination saves a lot of time for the plant breeders. Studies conclude that random DH's are comparable to the selected lines in pedigree inbreeding. The other advantages include development of large number of homozygous lines, efficient genetic analysis and development of markers for useful traits in much less time. More specific benefits include the possibility of seed propagation as an alternative to vegetative multiplication in ornamentals, and in species such as trees in which long life cycles and inbreeding depression preclude traditional breeding methods, doubled haploidy provides new alternatives.

DISADVANTAGES OF DHS

The main disadvantage with the DH population is that selection cannot be imposed on the population. But in conventional breeding selection can be practised for several generations: thereby desirable characters can be improved in the population.

In haploids produced from anther culture, it is observed that some plants are aneuploids and some are mixed haploid-diploid types. Another disadvantage associated with the double haploidy is the cost involved in establishing tissue culture and growth facilities. The over-usage of doubled haploidy may reduce genetic variation in breeding germplasm. Hence one has to take several factors into consideration before deploying doubled haploidy in breeding programmes.

Conclusions

Technological advances have now provided DH protocols for most plant genera. The number of species amenable to doubled haploidy has reached a staggering 250 in just a few decades. Response efficiency has also improved with gradual removal of species from recalcitrant category. Hence it will provide greater efficiency of plant breeding.

REFERENCES

1. Jain, S. Mohan, S. K. Sopory, and R. E. Veilleux. 1996. In vitro haploid production in higher plants. Dordrecht: Kluwer Academic Publishers. p.317.
2. Maluszynski et al., 2003.
3. B. Barnabás; B. Obert; G. Kovács (1999). "Colchicine, an efficient genome-doubling agent for maize (Zea mays L.) microspores cultured in anthero". Plant Cell Reports. 18 (10): 858–862. doi:10.1007/s002990050674.
4. Winzeler et al., 1987.
5. Forster and Thomas, 2003
6. Thomas et al., 2003.
7. Ardiel et al., 2002; William et al., 2002; Yi et al., 1998.
8. Thomas et al., 1984.
9. Schon et al., 1990.
10. RCSLs, Paterson et al., 1990.
11. STAIRS, Kearsey 2002.
12. Thomas et al., 2000.
13. Wang et al., 2001.
14. International Symposium on Genetic Manipulation in Crops. 1988. Genetic manipulation in crops proceedings of the International Symposium on Genetic Manipulation in Crops, the 3rd International Symposium on Haploidy, the 1st International Symposium on Somatic Cell Genetics in Crops, Beijing, October 1984. Natural resources and the environment series, v. 22. (London: Published for the International Rice Research Institute and Academia Sinica by Cassell Tycooly), p.318.
15. Immonen and Anttila, 1996.
16. Friedt et al., 1986; Winzeler et al., 1987.

CHAPTER 5

Genetic Engineering and Plants

Chapter Contents

- **Introduction**
- **Definition**
- **Process**
- **Applications**

Genetic engineering is the process of manually adding new DNA to an organism. The goal is to add one or more new traits that are not already found in that organism. Examples of genetically engineered (transgenic) organisms currently on the market include plants with resistance to some insects, plants that can tolerate herbicides, and crops with modified oil content.

INTRODUCTION

Genetic engineering, also called genetic modification, is the direct manipulation of an organism's genome using biotechnology. It is a set of technologies used to change the genetic makeup of cells, including the transfer of genes within and across species boundaries to produce improved or novel organisms. New DNA may be inserted in the host genome by first isolating and copying the genetic material of interest using molecular cloning methods to generate a DNA sequence, or by synthesizing the DNA, and then inserting this construct into the host organism. Genes may be removed, or "knocked out", using a nuclease. Gene targeting is a different technique that uses homologous recombination to change an endogenous gene, and can be used to delete a gene, remove exons, add a gene, or introduce point mutations.

An organism that is generated through genetic engineering is considered to be a genetically modified organism (GMO). The first GMOs were bacteria generated in 1973 and GM mice in 1974. Insulin-producing bacteria were commercialized in 1982 and genetically modified food has been sold since 1994. GloFish, the first GMO designed as a pet, was first sold in the United States in December 2003.

Genetic engineering techniques have been applied in numerous fields including research, agriculture, industrial biotechnology, and medicine. Enzymes used in laundry detergent and medicines such as insulin and human growth hormone are now manufactured in GM cells, experimental GM cell lines and GM animals such as mice or zebrafish are being used for research purposes, and genetically modified crops have been commercialized.

DEFINITION

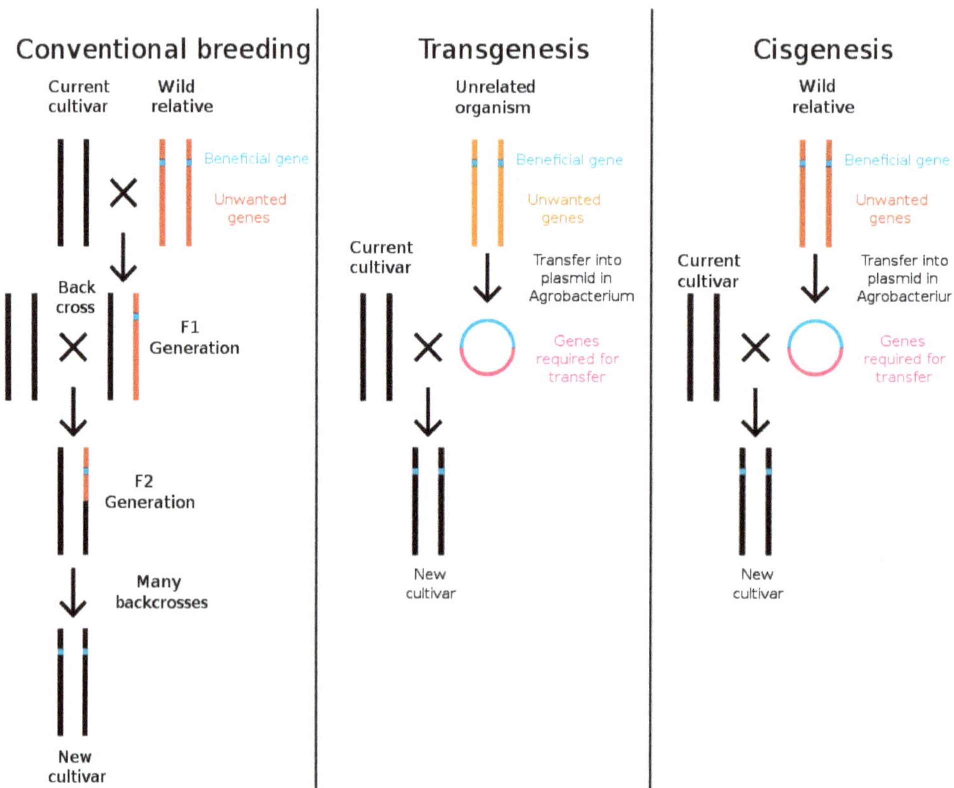

Comparison of conventional plant breeding with transgenic and cisgenic genetic modification.

Genetic engineering alters the genetic make-up of an organism using techniques that remove heritable material or that introduce DNA prepared outside the organism either directly into the host or into a cell that is then fused or hybridized with the host. This involves using recombinant nucleic acid (DNA or RNA) techniques to form new combinations of heritable genetic material followed by the incorporation of that material either indirectly through a vector system or directly through micro-injection, macro-injection and micro-encapsulation techniques.

Genetic engineering does not normally include traditional animal and plant breeding, in vitro fertilisation, induction of polyploidy, mutagenesis and cell fusion techniques that do not use recombinant nucleic acids or a genetically modified organism in the process. However the European Commission has also defined genetic engineering broadly as including selective breeding and other means of artificial selection. Cloning and stem cell research, although not considered genetic engineering, are closely related and genetic engineering can be used within them. Synthetic biology is an emerging discipline that takes genetic engineering a step further by introducing artificially synthesized material from raw materials into an organism.

If genetic material from another species is added to the host, the resulting organism is called transgenic. If genetic material from the same species or a species that can naturally breed with the host is used the resulting organism is called cisgenic. Genetic engineering can also be used to remove genetic material from the target organism, creating a gene knockout organism. In Europe genetic modification is synonymous with genetic engineering while within the United States of America it can also refer to conventional breeding methods. The Canadian regulatory system is based on whether a product has novel features regardless of method of origin. In other words, a product is regulated as genetically modified if it carries some trait not previously found in the species whether it was generated using traditional breeding methods (e.g., selective breeding, cell fusion, mutation breeding) or genetic engineering. Within the scientific community, the term *genetic engineering* is not commonly used; more specific terms such as *transgenic* are preferred.

Genetically Modified Organisms

Plants, animals or micro organisms that have changed through genetic engineering are termed genetically modified organisms or GMOs.Bacteria were the first organisms to be genetically modified. Plasmid DNA containing new genes can be inserted into the bacterial cell and the bacteria will then express those genes. These genes can code for medicines or enzymes that process food and other substrates. Plants have been modified for insect protection, herbicide resistance, virus resistance, enhanced nutrition, tolerance to environmental pressures and the production of edible vaccines. Most commercialised GMOs are insect resistant and/or herbicide tolerant crop plants. Genetically modified animals have been used for research, model animals and the production of agricultural or pharmaceutical products.

The genetically modified animals include animals with genes knocked out, increased susceptibility to disease, hormones for extra growth and the ability to express proteins in their milk.

History of Genetic Engineering

Genetic modification caused by human activity has been occurring since around 12,000 BC, when humans first began to domesticate organisms. Genetic engineering as the direct transfer of DNA from one organism to another was first accomplished by Herbert Boyer and Stanley Cohen in 1972. The first genetically modified animal was a mouse created in 1974 by Rudolf Jaenisch. In 1983 an antibiotic resistant gene was inserted into tobacco, leading to the first genetically engineered plant. Advances followed that allowed scientists to manipulate and add genes to a variety of different organism and induce a range of different effects.

In 1976 the technology was commercialized, with the advent of genetically modified bacteria that produced somatostatin, followed by insulin in 1978. Plants were first commercialized with virus resistant tobacco released in China in 1992. The first genetically modified food was the Flavr Savr tomato marketed in 1994. By 2010, 29 countries had planted commercialized biotech crops. In 2000 a paper published in *Science* introduced golden rice, the first food developed with increased nutrient value.

Herbert Boyer (pictured) and Stanley Cohen created the first genetically modified organism in 1972

Agriculture

DNA studies suggested that the dog most likely arose from a common ancestor with the grey wolf.

Genetic engineering is the direct manipulation of an organism's genome using certain biotechnology techniques that have only existed since the 1970s. Human directed genetic manipulation was occurring much earlier, beginning with the domestication of plants and animals through artificial selection. The dog is believed to be the first animal domesticated, possibly arising from a common ancestor of the grey wolf,with archeological evidence dating to about 12,000 BC. Other carnivores domesticated in prehistoric times include the cat, which cohabited with human 9,500 years ago. Archeological evidence suggests sheep, cattle, pigs and goats were domesticated between 9 000 BC and 8 000 BC in the Fertile Crescent.

The first evidence of plant domestication comes from emmer and einkorn wheat found in pre-Pottery Neolithic A villages in Southwest Asia dated about 10,500 to 10,100 BC. The Fertile Crescent of Western Asia, Egypt, and India were sites of the earliest planned sowing and harvesting of plants that had previously been gathered in the wild. Independent development of agriculture occurred in northern and southern China, Africa's Sahel, New Guinea and several regions of the Americas. The eight Neolithic founder crops (emmer wheat, einkorn wheat, barley, peas, lentils, bitter vetch, chick peas and flax) had all appeared by about 7000 BC. Horticulture first appears in the Levant during the Chalcolithic period about 6 800 to 6,300 BC. Due to the soft tissues, archeological evidence for early vegetables is scarce. The earliest vegetable remains have been found in Egyptian caves that date back to the 2nd millennium BC.

Selective breeding of domesticated plants was once the main way early farmers shaped organisms to suit their needs. Charles Darwin described three types of selection:

methodical selection, wherein humans deliberately select for particular characteristics; unconscious selection, wherein a characteristic is selected simply because it is desirable; and natural selection, wherein a trait that helps an organism survive better is passed on. Early breeding relied on unconscious and natural selection. The introduction of methodical selection is unknown. Common characteristics that were bred into domesticated plants include grains that did not shatter to allow easier harvesting, uniform ripening, shorter lifespans that translate to faster growing, loss of toxic compounds, and productivity. Some plants, like the Banana, were able to be propagated by vegetative cloning. Offspring often did not contain seeds, and therefore sterile. However, these offspring were usually juicier and larger. Propagation through cloning allows these mutant varieties to be cultivated despite their lack of seeds.

Hybridization was another way that rapid changes in plant's makeup were introduced. It often increased vigor in plants, and combined desirable traits together. Hybridization most likely first occurred when humans first grew similar, yet slightly different plants in close proximity. Triticum aestivum, wheat used in baking bread, is an allopolyploid. Its creation is the result of two separate hybridization events.

Grafting can transfer chloroplasts (specialised DNA in plants that can conduct photosynthesis), mitichondrial DNA and the entire cell nucleus containing the genome to potentially make a new species making grafting a form of natural genetic engineering.

X-rays were first used to deliberately mutate plants in 1927. Between 1927 and 2007, more than 2,540 genetically mutated plant varieties had been produced using x-rays.

Genetics

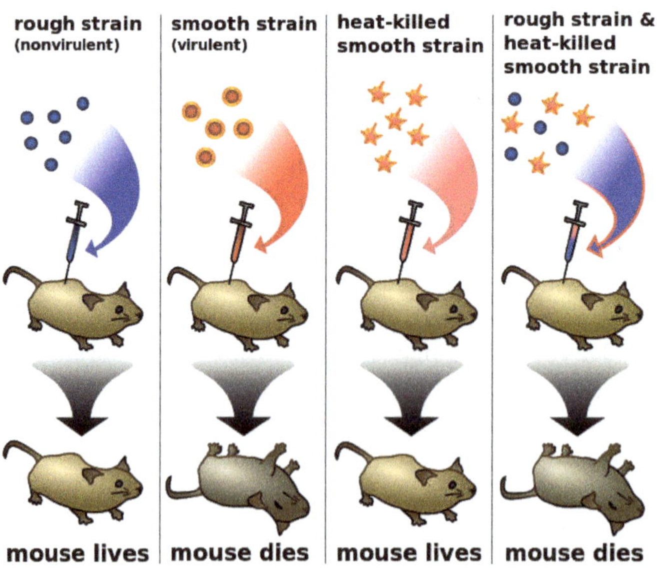

Griffith proved the existence of a "transforming principle", which Avery, MacLeod and McCarty later showed to be DNA

Various genetic discoveries have been essential in the development of genetic engineering. Genetic inheritance was first discovered by Gregor Mendel in 1865 following experiments crossing peas. Although largely ignored for 34 years he provided the first evidence of hereditary segregation and independent assortment. In 1889 Hugo de Vries came up with the name "(pan)gene" after postulating that particles are responsible for inheritance of characteristics and the term "genetics" was coined by William Bateson in 1905. In 1928 Frederick Griffith proved the existence of a "transforming principle" involved in inheritance, which Avery, MacLeod and McCarty later (1944) identified as DNA. Edward Lawrie Tatum and George Wells Beadle developed the central dogma that genes code for proteins in 1941. The double helix structure of DNA was identified by James Watson and Francis Crick in 1953.

As well as discovering how DNA works, tools had to be developed that allowed it to be manipulated. In 1970 Hamilton Smiths lab discovered restriction enzymes that allowed DNA to be cut at specific places and separated out on an electrophoresis gel. This enabled scientists to isolate genes from an organism's genome. DNA ligases, that join broken DNA together, had been discovered earlier in 1967 and by combining the two enzymes it was possible to "cut and paste" DNA sequences to create recombinant DNA. Plasmids, discovered in 1952,became important tools for transferring information between cells and replicating DNA sequences. Frederick Sanger developed a method for sequencing DNA in 1977, greatly increasing the genetic information available to researchers. Polymerase chain reaction (PCR), developed by Kary Mullis in 1983, allowed small sections of DNA to be amplified and aided identification and isolation of genetic material.

As well as manipulating the DNA, techniques had to be developed for its insertion (known as transformation) into an organism's genome. Griffiths experiment had already shown that some bacteria had the ability to naturally take up and express foreign DNA. Artificial competence was induced in *Escherichia coli* in 1970 when Morton Mandel and Akiko Higa showed that it could take up bacteriophage λ after treatment with calcium chloride solution ($CaCl_2$). Two years later, Stanley Cohen showed that $CaCl_2$ treatment was also effective for uptake of plasmid DNA. Transformation using electroporation was developed in the late 1980s, increasing the efficiency and bacterial range. In 1907 a bacterium that caused plant tumors, *Agrobacterium tumefaciens*, was discovered and in the early 1970s the tumor inducing agent was found to be a DNA plasmid called the Ti plasmid. By removing the genes in the plasmid that caused the tumor and adding in novel genes researchers were able to infect plants with *A. tumefaciens* and let the bacteria insert their chosen DNA into the genomes of the plants.

Early Genetically Modified Organisms

In 1972 Paul Berg utilised restriction enzymes and DNA ligases to create the first recombinant DNA molecules. He combined DNA from the monkey virus SV40 with that of the lambda virus. Herbert Boyer and Stanley Norman Cohen took Berg's work a step further and introduced recombinant DNA into a bacterial cell. Cohen was researching plasmids, while Boyers work involved restriction enzymes. They recognised the complementary nature of their work and teamed up in 1972. Together they found a restriction enzyme that cut the pSC101 plasmid at a single point and were able to insert and ligate a gene that conferred resistance to the kanamycin antibiotic into the gap. Cohen had previously devised a method where bacteria could be induced to take up a plasmid and using this they were able to create a bacteria that survived in the presence of the kanamycin. This represented the first genetically modified organism. They repeated experiments showing that other genes could be expressed in bacteria, including one from the toad Xenopus laevis, the first cross kingdom transformation.

Paul Berg created the first recombinant DNA molecules in 1972.

In 1974 Rudolf Jaenisch created a transgenic mouse by introducing foreign DNA into its embryo, making it the world's first transgenic animal. Jaenisch was studying mammalian cells infected with simian virus 40 (SV40) when he happened to read a paper from Beatrice Mintz describing the generation of chimera mice. He took his SV40 samples to Mintz's lab and injected them into early mouse embryos expecting tumours to develop. The mice appeared normal, but after using radioactive probes he discovered that the virus had integrated itself into the mice genome. However the mice did not pass the transgene to their offspring. In 1981 the laboratories of Frank Ruddle, Frank Constantini and Elizabeth Lacy injected purified DNA into a single-cell mouse embryo and showed transmission of the genetic material to subsequent generations.

In 1974 Rudolf Jaenisch created the first GM animal.

The first genetically engineered plant was tobacco, reported in 1983. It was developed by Michael W. Bevan, Richard B. Flavell and Mary-Dell Chilton by creating a chimeric gene that joined an antibiotic resistant gene to the T1 plasmid from *Agrobacterium*. The tobacco was infected with *Agrobacterium* transformed with this plasmid resulting in the chimeric gene being inserted into the plant. Through tissue culture techniques a single tobacco cell was selected that contained the gene and a new plant grown from it.

Regulation

The development of genetic engineering technology led to concerns in the scientific community about potential risks. The development of a regulatory framework concerning genetic engineering began in 1975, at Asilomar, California. The Asilomar meeting recommended a set of guidelines regarding the cautious use of recombinant technology and any products resulting from that technology. The Asilomar recommendations were voluntary, but in 1976 the US National Institute of Health (NIH) formed a recombinant DNA advisory committee. This was followed by other regulatory offices (the United States Department of Agriculture (USDA), Environmental Protection Agency (EPA) and Food and Drug Administration (FDA), effectively making all recombinant DNA research tightly regulated in the USA.

In 1982 the Organization for Economic Co-operation and Development (OECD) released a report into the potential hazards of releasing genetically modified organisms into the environment as the first transgenic plants were being developed. As the technology improved and genetically organisms moved from model organisms to potential commercial products the USA established a committee at the Office of Science and Technology (OSTP) to develop mechanisms to regulate the developing technology. In 1986 the OSTP assigned regulatory approval of genetically modified plants in the US to the USDA, FDA and EPA. In the late 1980s and early 1990s, guidance on assessing the safety of genetically engineered plants and food emerged from organizations including the FAO and WHO.

The European Union first introduced laws requiring GMO's to be labelled in 1997. In 2013 Connecticut became the first state to enact a labeling law in the USA, although it would not take effect until other states followed suit.

Research and Medicine

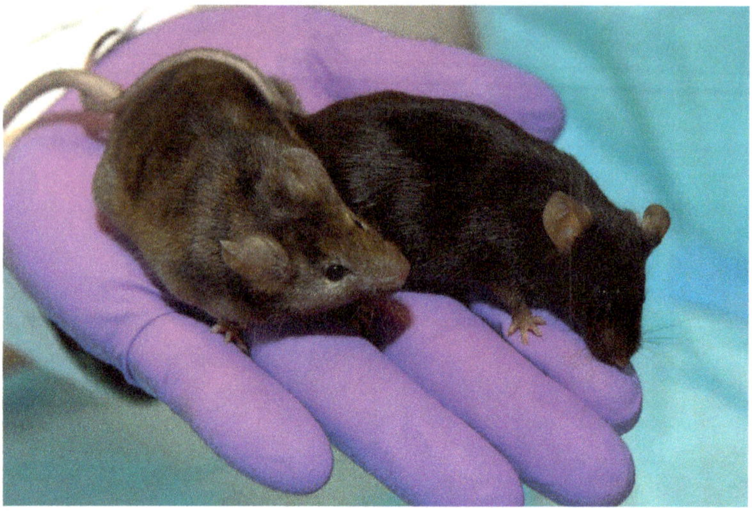

A laboratory mouse in which a gene affecting hair growth has been knocked out (left), is shown next to a normal lab mouse.

The ability to insert, alter or remove genes in model organisms allowed scientists to study the genetic elements of human diseases. Genetically modified mice were created in 1984 that carried cloned oncogenes that predisposed them to developing cancer. The technology has also been used to generate mice with genes knocked out. The first recorded knockout mouse was created by Mario R. Capecchi, Martin Evans and Oliver Smithies in 1989. In 1992 oncomice with tumor suppressor genes knocked out were generated. Creating Knockout rats is much harder and only became possible in 2003.

After the discovery of microRNA in 1993, RNA interference (RNAi) has been used to silence an organism's genes. By modifying an organism to express microRNA targeted to its endogenous genes, researchers have been able to knockout or partially reduce gene function in a range of species. The ability to partially reduce gene function has allowed the study of genes that are lethal when completely knocked out. Other advantages of using RNAi include the availability of inducible and tissue specific knockout. In 2007 microRNA targeted to insect and nematode genes was expressed in plants, leading to suppression when they fed on the transgenic plant, potentially creating a new way to control pests. Targeting endogenous microRNA expression has allowed further fine tuning of gene expression, supplementing the more traditional gene knock out approach.

Genetic engineering has been used to produce proteins derived from humans and other sources in organisms that normally cannot synthesize these proteins. Human insulin-synthesising bacteria were developed in 1979 and were first used as a treatment in 1982. In 1988 the first human antibodies were produced in plants. In 2000 Vitamin A-enriched golden rice, was the first food with increased nutrient value.

In 2014 a bacteria was developed that replicated a plasmid containing an unnatural base pair. This required altering the bacterium so it could import the unnatural nucleotides and then efficiently replicate them. The plasmid retained the unnatural base pairs when it doubled an estimated 99.4% of the time. This is the first organism engineered to use an expanded genetic alphabet.

In 2015 CRISPR and TALENs was used to modify plant genomes. Chinese labs used it to create a fungus-resistant wheat and boost rice yields, while a U.K. group used it to tweak a barley gene that could help produce drought-resistant varieties. When used to precisely remove material from DNA without adding genes from other species, the result is not subject the lengthy and expensive regulatory process associated with GMOs. While CRISPR may use foreign DNA to aid the editing process, the second generation of edited plants contain none of that DNA. Researchers celebrated the acceleration because it may allow them to "keep up" with rapidly evolving pathogens. The U.S. Department of Agriculture stated that some examples of gene-edited corn, potatoes and soybeans are not subject to existing regulations. As of 2016 other review bodies had yet to make statements.

Commercialisation

In 1976 Genentech, the first genetic engineering company was founded by Herbert Boyer and Robert Swanson and a year later and the company produced a human protein (somatostatin) in *E.coli*. Genentech announced the production of genetically engineered human insulin in 1978. In 1980 the U.S. Supreme Court in the Diamond v. Chakrabarty case ruled that genetically altered life could be patented. The insulin produced by bacteria, branded humulin, was approved for release by the Food and Drug Administration in 1982.

In 1983 a biotech company, Advanced Genetic Sciences (AGS) applied for U.S. government authorization to perform field tests with the ice-minus strain of *P. syringae* to protect crops from frost, but environmental groups and protestors delayed the field tests for four years with legal challenges. In 1987 the ice-minus strain of *P. syringae* became the first genetically modified organism (GMO) to be released into the environment when a strawberry field and a potato field in California were sprayed with it. Both test fields were attacked by activist groups the night before the tests occurred: "The world's first trial site attracted the world's first field trasher".

The first genetically modified crop plant was produced in 1982, an antibiotic-resistant tobacco plant. The first field trials of genetically engineered plants occurred in France and the USA in 1986, tobacco plants were engineered to be resistant to herbicides. In

1987 Plant Genetic Systems, founded by Marc Van Montagu and Jeff Schell, was the first company to genetically engineer insect-resistant plants by incorporating genes that produced insecticidal proteins from Bacillus thuringiensis (Bt) into tobacco.

Genetically modified microbial enzymes were the first application of genetically modified organisms in food production and were approved in 1988 by the US Food and Drug Administration. In the early 1990s, recombinant chymosin was approved for use in several countries. Cheese had typically been made using the enzyme complex rennet that had been extracted from cows' stomach lining. Scientists modified bacteria to produce chymosin, which was also able to clot milk, resulting in cheese curds. The People's Republic of China was the first country to commercialize transgenic plants, introducing a virus-resistant tobacco in 1992. In 1994 Calgene attained approval to commercially release the Flavr Savr tomato, a tomato engineered to have a longer shelf life. Also in 1994, the European Union approved tobacco engineered to be resistant to the herbicide bromoxynil, making it the first genetically engineered crop commercialized in Europe. In 1995 Bt Potato was approved safe by the Environmental Protection Agency, after having been approved by the FDA, making it the first pesticide producing crop to be approved in the USA. In 1996 a total of 35 approvals had been granted to commercially grow 8 transgenic crops and one flower crop (carnation), with 8 different traits in 6 countries plus the EU.

By 2010, 29 countries had planted commercialized biotech crops and a further 31 countries had granted regulatory approval for transgenic crops to be imported. In 2013 Robert Fraley (Monsanto's executive vice president and chief technology officer), Marc Van Montagu and Mary-Dell Chilton were awarded the World Food Prize for improving the "quality, quantity or availability" of food in the world.

The first genetically modified animal to be commercialised was the GloFish, a Zebra fish with a fluorescent gene added that allows it to glow in the dark under ultraviolet light. The first genetically modified animal to be approved for food use was AquAdvantage salmon in 2015. The salmon were transformed with a growth hormone-regulating gene from a Pacific Chinook salmon and a promoter from an ocean pout enabling it to grow year-round instead of only during spring and summer.

Opposition

Opposition and support for the use of genetic engineering has existed since the technology was developed. After Arpad Pusztai went public with research he was conducting in 1998 the public opposition to genetically modified food increased. Opposition continued following controversial and publicly debated papers published in 1999 and 2013 that claimed negative environmental and health impacts from genetically modified crops.

PROCESS

The first step is to choose and isolate the gene that will be inserted into the genetically modified organism. The gene can be isolated using restriction enzymes to cut DNA into fragments and gel electrophoresis to separate them out according to length. Polymerase chain reaction (PCR) can also be used to amplify up a gene segment, which can then be isolated through gel electrophoresis. If the chosen gene or the donor organism's genome has been well studied it may be present in a genetic library. If the DNA sequence is known, but no copies of the gene are available, it can be artificially synthesized.

The gene to be inserted into the genetically modified organism must be combined with other genetic elements in order for it to work properly. The gene can also be modified at this stage for better expression or effectiveness. As well as the gene to be inserted most constructs contain a promoter and terminator region as well as a selectable marker gene. The promoter region initiates transcription of the gene and can be used to control the location and level of gene expression, while the terminator region ends transcription. The selectable marker, which in most cases confers antibiotic resistance to the organism it is expressed in, is needed to determine which cells are transformed with the new gene. The constructs are made using recombinant DNA techniques, such as restriction digests, ligations and molecular cloning. The manipulation of the DNA generally occurs within a plasmid.

The most common form of genetic engineering involves inserting new genetic material randomly within the host genome. Other techniques allow new genetic material to be inserted at a specific location in the host genome or generate mutations at desired genomic loci capable of knocking out endogenous genes. The technique of gene targeting uses homologous recombination to target desired changes to a specific endogenous gene. This tends to occur at a relatively low frequency in plants and animals and generally requires the use of selectable markers. The frequency of gene targeting can be greatly enhanced with the use of engineered nucleases such as zinc finger nucleases,engineered homing endonucleases, or nucleases created from TAL effectors.

In addition to enhancing gene targeting, engineered nucleases can also be used to introduce mutations at endogenous genes that generate a gene knockout.

Transformation

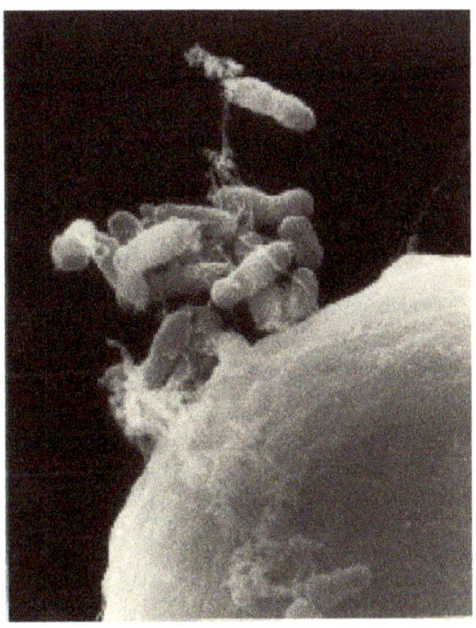

A. tumefaciens attaching itself to a carrot cell

Only about 1% of bacteria are naturally capable of taking up foreign DNA. However, this ability can be induced in other bacteria via stress (e.g. thermal or electric shock), thereby increasing the cell membrane's permeability to DNA; up-taken DNA can either integrate with the genome or exist as extrachromosomal DNA. DNA is generally inserted into animal cells using microinjection, where it can be injected through the cell's nuclear envelope directly into the nucleus or through the use of viral vectors. In plants the DNA is generally inserted using *Agrobacterium*-mediated recombination or biolistics.

In *Agrobacterium*-mediated recombination, the plasmid construct contains T-DNA, DNA which is responsible for insertion of the DNA into the host plants genome. This plasmid is transformed into *Agrobacterium* containing no plasmids prior to infecting the plant cells. The *Agrobacterium* will then naturally insert the genetic material into the plant cells. In biolistics transformation particles of gold or tungsten are coated with DNA and then shot into young plant cells or plant embryos. Some genetic material will enter the cells and transform them. This method can be used on plants that are not susceptible to *Agrobacterium* infection and also allows transformation of plant plastids. Another transformation method for plant and animal cells is electroporation. Electroporation involves subjecting the plant or animal cell to an electric shock, which can make the cell membrane permeable to plasmid DNA. In some cases the electroporated cells will incorporate the DNA into their genome. Due to the damage caused to the cells and DNA the transformation efficiency of biolistics and electroporation is lower than agrobacterial mediated transformation and microinjection.

As often only a single cell is transformed with genetic material the organism must be regenerated from that single cell. As bacteria consist of a single cell and reproduce clonally regeneration is not necessary. In plants this is accomplished through the use of tissue culture. Each plant species has different requirements for successful regeneration through tissue culture. If successful an adult plant is produced that contains the transgene in every cell. In animals it is necessary to ensure that the inserted DNA is present in the embryonic stem cells. Selectable markers are used to easily differentiate transformed from untransformed cells. These markers are usually present in the transgenic organism, although a number of strategies have been developed that can remove the selectable marker from the mature transgenic plant. When the offspring is produced they can be screened for the presence of the gene. All offspring from the first generation will be heterozygous for the inserted gene and must be mated together to produce a homozygous animal.

Further testing uses PCR, Southern hybridization, and DNA sequencing is conducted to confirm that an organism contains the new gene. These tests can also confirm the chromosomal location and copy number of the inserted gene. The presence of the gene does not guarantee it will be expressed at appropriate levels in the target tissue so methods that look for and measure the gene products (RNA and protein) are also used. These include northern hybridization, quantitative RT-PCR, Western blot, immunofluorescence, ELISA and phenotypic analysis. For stable transformation the gene should be passed to the offspring in a Mendelian inheritance pattern, so the organism's offspring are also studied.

Genome Editing

Genome editing is a type of genetic engineering in which DNA is inserted, replaced, or removed from a genome using artificially engineered nucleases, or "molecular scissors." The nucleases create specific double-stranded breaks (DSBs) at desired locations in the genome, and harness the cell's endogenous mechanisms to repair the induced break by natural processes of homologous recombination (HR) and nonhomologous end-joining (NHEJ). There are currently four families of engineered nucleases: meganucleases, zinc finger nucleases (ZFNs), transcription activator-like effector nucleases (TALENs), and the Cas9-guideRNA system (adapted from the CRISPR prokarotic immune system). In contrast to artificial genome editing natural genome editing occurs through viral and sub-viral agents competent in identification of genetic syntax structures for insertion/deletion processes with the result of conserved selection processes.

APPLICATIONS

Genetic engineering has applications in medicine, research, industry and agriculture and can be used on a wide range of plants, animals and micro organisms.

Medicine

In medicine, genetic engineering has been used in manufacturing drugs, creating model animals, conducting laboratory research, and in gene therapy.

Manufacturing

Genetic engineering is used to mass-produce insulin, human growth hormones, follistim (for treating infertility), human albumin, monoclonal antibodies, antihemophilic factors, vaccines and many other drugs. Mouse hybridomas, cells fused together to create monoclonal antibodies, have been humanised through genetic engineering to create human monoclonal antibodies. Genetically engineered viruses are being developed that can still confer immunity, but lack the infectious sequences.

Research

Genetic engineering is used to create animal models of human diseases. Genetically modified mice are the most common genetically engineered animal model. They have been used to study and model cancer (the oncomouse), obesity, heart disease, diabetes, arthritis, substance abuse, anxiety, aging and Parkinson disease. Potential cures can be tested against these mouse models. Also genetically modified pigs have been bred with the aim of increasing the success of pig to human organ transplantation.

Gene Therapy

Gene therapy is the genetic engineering of humans, generally by replacing defective genes with effective ones. This can occur in somatic tissue or germline tissue.

Somatic gene therapy has been studied in clinical research in several diseases, including X-linked SCID, chronic lymphocytic leukemia (CLL), and Parkinson's disease. In 2012, Glybera became the first gene therapy treatment to be approved for clinical use in either Europe or the United States after its endorsement by the European Commission.

With regard to germline gene therapy, the scientific community has been opposed to attempts to alter genes in humans in inheritable ways using biotechnology since the technology was first introduced, and the caution has continued as the technology has progressed. With the advent of new techniques like CRISPR, in March 2015 scientists urged a worldwide ban on clinical use of gene editing technologies to edit the human genome in a way that can be inherited. In April 2015, Chinese researchers sparked controversy when they reported results of basic research experiments in which they edited the DNA of non-viable human embryos using CRISPR. In December 2015, scientists of major world academies called for a moratorium on inheritable human genome edits, including those related to CRISPR-Cas9 technologies.

There are also ethical concerns should the technology be used not just for treatment, but for enhancement, modification or alteration of a human beings' appearance, adaptability, intelligence, character or behavior. The distinction between cure and enhancement can also be difficult to establish. Transhumanists consider the enhancement of humans desirable.

Research

Knockout mice

Genetic engineering is an important tool for natural scientists. Genes and other genetic information from a wide range of organisms are transformed into bacteria for storage and modification, creating genetically modified bacteria in the process. Bacteria are cheap, easy to grow, clonal, multiply quickly, relatively easy to transform and can be stored at -80 °C almost indefinitely. Once a gene is isolated it can be stored inside the bacteria providing an unlimited supply for research.

Organisms are genetically engineered to discover the functions of certain genes. This could be the effect on the phenotype of the organism, where the gene is expressed or what other genes it interacts with. These experiments generally involve loss of function, gain of function, tracking and expression.

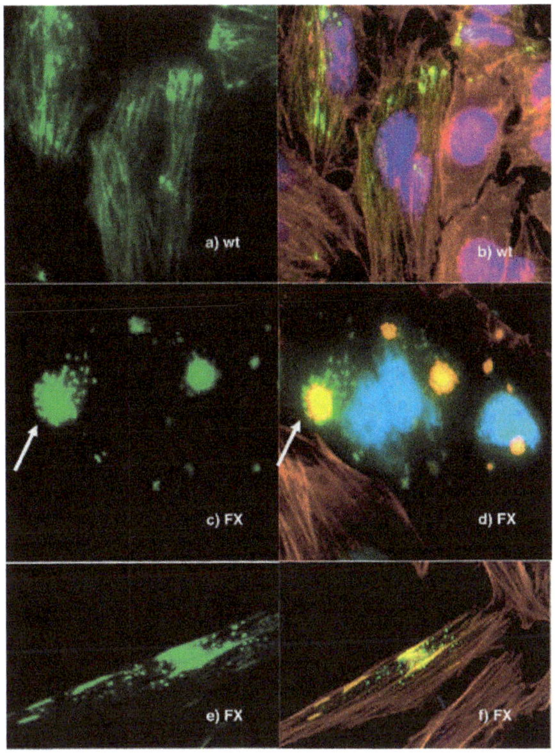

Human cells in which some proteins are fused with green fluorescent protein to allow them to be visualised

- **Loss of function experiments**, such as in a gene knockout experiment, in which an organism is engineered to lack the activity of one or more genes. A knockout experiment involves the creation and manipulation of a DNA construct *in vitro*, which, in a simple knockout, consists of a copy of the desired gene, which has been altered such that it is non-functional. Embryonic stem cells incorporate the altered gene, which replaces the already present functional copy. These stem cells are injected into blastocysts, which are implanted into surrogate mothers. This allows the experimenter to analyze the defects caused by this mutation and thereby determine the role of particular genes. It is used especially frequently in developmental biology. Another method, useful in organisms such as Drosophila (fruit fly), is to induce mutations in a large population and then screen the progeny for the desired mutation. A similar process can be used in both plants and prokaryotes. Loss of function tells whether or not a protein is required for a function, but does not always mean it's sufficient, especially if a function requires multiple proteins and is lost if one protein is missing.
- **Gain of function experiments**, the logical counterpart of knockouts. These are sometimes performed in conjunction with knockout experiments to more finely establish the function of the desired gene. The process is much the same as that in knockout engineering, except that the construct is designed to increase the function of the gene, usually by providing extra copies of the gene or

inducing synthesis of the protein more frequently. Gain of function is used to tell whether or not a protein is sufficient for a function, but does not always mean it's required, especially when dealing with genetic or functional redundancy.

- **Tracking experiments**, which seek to gain information about the localization and interaction of the desired protein. One way to do this is to replace the wild-type gene with a 'fusion' gene, which is a juxtaposition of the wild-type gene with a reporting element such as green fluorescent protein (GFP) that will allow easy visualization of the products of the genetic modification. While this is a useful technique, the manipulation can destroy the function of the gene, creating secondary effects and possibly calling into question the results of the experiment. More sophisticated techniques are now in development that can track protein products without mitigating their function, such as the addition of small sequences that will serve as binding motifs to monoclonal antibodies.
- **Expression studies** aim to discover where and when specific proteins are produced. In these experiments, the DNA sequence before the DNA that codes for a protein, known as a gene's promoter, is reintroduced into an organism with the protein coding region replaced by a reporter gene such as GFP or an enzyme that catalyzes the production of a dye. Thus the time and place where a particular protein is produced can be observed. Expression studies can be taken a step further by altering the promoter to find which pieces are crucial for the proper expression of the gene and are actually bound by transcription factor proteins; this process is known as promoter bashing.

Industrial

Using genetic engineering techniques, one can transform microorganisms such as bacteria or yeast, or transform cells from multicellular organisms such as insects or mammals, with a gene coding for a useful protein, such as an enzyme, so that the transformed organism will overexpress the desired protein. One can manufacture mass quantities of the protein by growing the transformed organism in bioreactor equipment using techniques of industrial fermentation, and then purifying the protein. Some genes do not work well in bacteria, so yeast, insect cells, or mammalians cells, each a eukaryote, can also be used. These techniques are used to produce medicines such as insulin, human growth hormone, and vaccines, supplements such as tryptophan, aid in the production of food (chymosin in cheese making) and fuels. Other applications involving genetically engineered bacteria being investigated involve making the bacteria perform tasks outside their natural cycle, such as making biofuels, cleaning up oil spills, carbon and other toxic waste and detecting arsenic in drinking water. Certain genetically modified microbes can also be used in biomining and bioremediation, due to their ability to extract heavy metals from their environment and incorporate them into compounds that are more easily recoverable.

Experimental, Lab Scale Industrial Applications

In materials science, a genetically modified virus has been used in an academic lab as a scaffold for assembling a more environmentally friendly lithium-ion battery.

Bacteria have been engineered to function as sensors by expressing a fluorescent protein under certain environmental conditions.

Agriculture

Bt-toxins present in peanut leaves (bottom image) protect it from extensive damage caused by European corn borerlarvae (top image).

One of the best-known and controversial applications of genetic engineering is the creation and use of genetically modified crops or genetically modified organisms, such as genetically modified fish, which are used to produce genetically modified food and materials with diverse uses. There are four main goals in generating genetically modified crops.

One goal, and the first to be realized commercially, is to provide protection from environmental threats, such as cold (in the case of Ice-minus bacteria), or pathogens, such as insects or viruses, and/or resistance to herbicides. There are also fungal and virus resistant crops developed or in development. They have been developed to make the insect and weed management of crops easier and can indirectly increase crop yield.

Another goal in generating GMOs is to modify the quality of produce by, for instance, increasing the nutritional value or providing more industrially useful qualities or quantities. The Amflora potato, for example, produces a more industrially useful blend of starches. Cows have been engineered to produce more protein in their milk to facilitate cheese production. Soybeans and canola have been genetically modified to produce more healthy oils.

Another goal consists of driving the GMO to produce materials that it does not normally make. One example is "pharming", which uses crops as bioreactors to produce vaccines, drug intermediates, or drug themselves; the useful product is purified from the harvest and then used in the standard pharmaceutical production process.Cows and goats have been engineered to express drugs and other proteins in their milk, and in 2009 the FDA approved a drug produced in goat milk.

Another goal in generating GMOs, is to directly improve yield by accelerating growth, or making the organism more hardy (for plants, by improving salt, cold or drought tolerance). Salmon have been genetically modified with growth hormones to increase their size.

The genetic engineering of agricultural crops can increase the growth rates and resistance to different diseases caused by pathogens and parasites. This is beneficial as it can greatly increase the production of food sources with the usage of fewer resources that would be required to host the world's growing populations. These modified crops would also reduce the usage of chemicals, such as fertilizers and pesticides, and therefore decrease the severity and frequency of the damages produced by these chemical pollution.

There is a scientific consensus that currently available food derived from GM crops poses no greater risk to human health than conventional food, but that each GM food needs to be tested on a case-by-case basis before introduction.Nonetheless, members of the public are much less likely than scientists to perceive GM foods as safe. The legal and regulatory status of GM foods varies by country, with some nations banning or restricting them, and others permitting them with widely differing degrees of regulation. Gene flow into related non-transgenic crops, off target effects on beneficial organisms and the impact on biodiversity are important environmental issues. Ethical concerns involve religious issues, corporate control of the food supply, intellectual property rights and the level of labeling needed on genetically modified products.

Conservation

Genetic engineering has potential applications in conservation and natural areas management. For example, gene transfer through viral vectors has been proposed as a means of controlling invasive species as well as vaccinating threatened fauna from disease. Transgenic trees have also been suggested as a way to confer resistance to pathogens in wild populations. With the increasing risks of maladaptation in organisms as a result of climate change and other perturbations, facilitated adaptation through gene tweaking could be one solution to reducing extinction risks. Applications of genetic engineering in conservation are thus far mostly theoretical and have yet to be put into practice. Further experimentation will be necessary to gauge the benefits and costs of such practices.

BioArt and Entertainment

Genetic engineering is also being used to create BioArt. Some bacteria have been genetically engineered to create black and white photographs.

Genetic engineering has also been used to create novelty items such as lavender-colored carnations, blue roses, and glowing fish.

Regulation

The regulation of genetic engineering concerns the approaches taken by governments to assess and manage the risks associated with the development and release of genetically modified crops. There are differences in the regulation of GM crops between countries, with some of the most marked differences occurring between the USA and Europe. Regulation varies in a given country depending on the intended use of the products of the genetic engineering. For example, a crop not intended for food use is generally not reviewed by authorities responsible for food safety. Starting in the late 1980s, guidance on assessing the safety of genetically engineered plants and food emerged from organizations including the FAO and WHO.

Controversy

Critics have objected to use of genetic engineering per se on several grounds, including ethical concerns, ecological concerns, and economic concerns raised by the fact GM techniques and GM organisms are subject to intellectual property law. GMOs also are involved in controversies over GM food with respect to whether food produced from GM crops is safe, whether it should be labeled, and whether GM crops are needed to address the world's food needs. See the genetically modified food controversies article for discussion of issues about GM crops and GM food. These controversies have led to litigation, international trade disputes, and protests, and to restrictive regulation of commercial products in some countries.

REFERENCES

1. "First transgenic pet, 'GloFish', sold to US public". PHG Foundation. 9 January 2004.
2. "Terms and Acronyms". U.S. Environmental Protection Agency online. Retrieved 16 July 2015.
3. Vert, Michel; Doi, Yoshiharu; Hellwich, Karl-Heinz; Hess, Michael; Hodge, Philip; Kubisa, Przemyslaw; Rinaudo, Marguerite; Schué, François (2012). "Terminology for biorelated polymers and applications (IUPAC Recommendations 2012)" (PDF). Pure and Applied Chemistry. 84 (2): 377–410. doi:10.1351/PAC-REC-10-12-04.
4. The European Parliament and the council of the European Union (12 March 2001). "Directive on the release of genetically modified organisms (GMOs) Directive

2001/18/EC ANNEX I A". Official Journal of the European Communities: 17.

5. Staff Economic Impacts of Genetically Modified Crops on the Agri-Food Sector; P. 42 Glossary - Term and Definitions The European Commission Directorate-General for Agriculture, "Genetic engineering: The manipulation of an organism's genetic endowment by introducing or eliminating specific genes through modern molecular biology techniques. A broad definition of genetic engineering also includes selective breeding and other means of artificial selection.", Retrieved 5 November 2012
6. Van Eenennaam, Alison. "Is Livestock Cloning Another Form of Genetic Engineering?" (PDF). agbiotech. Archived from the original (PDF) on 11 May 2011.
7. Suter, David M.; Dubois-Dauphin, Michel; Krause, Karl-Heinz (2006). "Genetic engineering of embryonic stem cells" (PDF). Swiss Med Wkly. 136 (27–28): 413–415. PMID 16897894.
8. Andrianantoandro, Ernesto; Basu, Subhayu; Kariga, David K.; Weiss, Ron (16 May 2006). "Synthetic biology: new engineering rules for an emerging discipline". Molecular Systems Biology. 2 (2006.0028): 2006.0028. doi:10.1038/msb4100073. PMC 1681505 . PMID 16738572.
9. Jacobsen, E.; Schouten, H. J. (2008). "Cisgenesis, a New Tool for Traditional Plant Breeding, Should be Exempted from the Regulation on Genetically Modified Organisms in a Step by Step Approach". Potato Research. 51: 75–88. doi:10.1007/s11540-008-9097-y.
10. Capecchi, Mario R. (2001). "Generating mice with targeted mutations". Nature Medicine. 7 (10): 1086–90. doi:10.1038/nm1001-1086. PMID 11590420.
11. Staff Biotechnology - Glossary of Agricultural Biotechnology TermsUnited States Department of Agriculture, "Genetic modification: The production of heritable improvements in plants or animals for specific uses, via either genetic engineering or other more traditional methods. Some countries other than the United States use this term to refer specifically to genetic engineering.", Retrieved 5 November 2012
12. Maryanski, James H. (19 October 1999). "Genetically Engineered Foods". Center for Food Safety and Applied Nutrition at the Food and Drug Administration.
13. Evans, Brent and Lupescu, Mihai (15 July 2012) Canada - Agricultural Biotechnology Annual – 2012 GAIN (Global Agricultural Information Network) report CA12029, United States Department of Agriculture, Foreifn Agricultural Service, Retrieved 5 November 2012
14. McHugen, Alan (14 September 2000). "Chapter 1: Hors-d'oeuvres and entrees/What is genetic modification? What are GMOs?". Pandora's Picnic Basket. Oxford University Press. ISBN 978-0198506744.
15. Staff (28 November 2005) Health Canada - The Regulation of Genetically Modified Food Glossary definition of Genetically Modified: "An organism, such as a plant, animal or bacterium, is considered genetically modified if its genetic material has been altered through any method, including conventional breeding.

A 'GMO' is a genetically modified organism.", Retrieved 5 November 2012

16. "What is genetic modification (GM)?". CSIRO.
17. "Genetic Modification of Bacteria". Annenberg Foundation.
18. Panesar, Pamit et al (2010) "Enzymes in Food Processing: Fundamentals and Potential Applications", Chapter 10, I K International Publishing House, ISBN 978-9380026336
19. "GM traits list". International Service for the Acquisition of Agri-Biotech Applications.
20. "ISAAA Brief 43-2011: Executive Summary". International Service for the Acquisition of Agri-Biotech Applications.
21. Connor, Steve (2 November 2007). "The mouse that shook the world". The Independent.
22. Root, Clive (2007). Domestication. Greenwood Publishing Groups.
23. Zohary, Daniel; Hopf, Maria; Weiss, Ehud (2012). Domestication of Plants in the Old World: The origin and spread of plants in the old world. Oxford University Press.
24. Stableford, Brian M. (2004). Historical dictionary of science fiction literature. p. 133. ISBN 9780810849389.
25. A, Hershey; Chase, M. (1952). "Independent functions of viral protein and nucleic acid in growth of bacteriophage" (PDF). J Gen Physiol. 36 (1): 39–56. doi:10.1085/jgp.36.1.39. PMC 2147348 . PMID 12981234.
26. "Genetic Engineering". Encyclopedia of Science Fiction. April 2, 2015.
27. Shiv Kant Prasad, Ajay Dash (2008). Modern Concepts in Nanotechnology, Volume 5. Discovery Publishing House. ISBN 9788183562966.
28. Jackson, DA; Symons, RH; Berg, P (1 October 1972). "Biochemical Method for Inserting New Genetic Information into DNA of Simian Virus 40: Circular SV40 DNA Molecules Containing Lambda Phage Genes and the Galactose Operon of Escherichia coli". PNAS. 69 (10): 2904–2909. Bibcode:1972PNAS...69.2904J. doi:10.1073/pnas.69.10.2904. PMC 389671 . PMID 4342968.
29. Arnold, Paul (2009). "History of Genetics: Genetic Engineering Timeline".
30. Cohen, Stanley N.; Chang, Annie C. Y. (1 May 1973). "Recircularization and Autonomous Replication of a Sheared R-Factor DNA Segment in Escherichia coli Transformants — PNAS". Pnas.org. Retrieved 17 July 2010.
31. Jaenisch, R. and Mintz, B. (1974) Simian virus 40 DNA sequences in DNA of healthy adult mice derived from preimplantation blastocysts injected with viral DNA. Proc. Natl. Acad. 71(4) 1250–1254
32. Berg P; Baltimore, D; Brenner, S; Roblin, RO; Singer, MF; et al. (1975). "Summary statement of the Asilomar Conference on recombinant DNA molecules" (PDF). Proc. Natl. Acad. Sci. U.S.A. 72 (6): 1981–4. Bibcode:1975PNAS...72.1981B. doi:10.1073/pnas.72.6.1981. PMC 432675 . PMID 806076.
33. NIH Guidelines for research involving recombinant DNA moleculesArchived

10 September 2012 at the Wayback Machine.

34. Goeddel, David; Kleid, Dennis G.; Bolivar, Francisco; Heyneker, Herbert L.; Yansura, Daniel G.; Crea, Roberto; Hirose, Tadaaki; Kraszewski, Adam; Itakura, Keiichi; Riggs, Arthur D. (January 1979). "Expression in Escherichia coli of chemically synthesized genes for human insulin" (PDF). PNAS. 76 (1): 106–110. Bibcode:1979PNAS...76..106G. doi:10.1073/pnas.76.1.106. PMC 382885 . PMID 85300.
35. US Supreme Court Cases from Justia & Oyez (16 June 1980). "Diamond V Chakrabarty". 447 (303). Supreme.justia.com. Retrieved 17 July 2010.
36. "Artificial Genes". TIME. 15 November 1982. Retrieved 17 July 2010.
37. H. Patricia Hynes. (1989) Biotechnology in agriculture: an analysis of selected technologies and policy in the United States. Reproductive and Genetic Engineering (2)1:39–49
38. Bratspies, Rebecca (2007). "Some Thoughts on the American Approach to Regulating Genetically Modified Organisms" (PDF). Kansas Journal of Law & Public Policy. 16 (3): 101–131.
39. BBC News 14 June 2002 GM crops: A bitter harvest?
40. Thomas H. Maugh II for the Los Angeles Times. 9 June 1987. Altered Bacterium Does Its Job : Frost Failed to Damage Sprayed Test Crop, Company Says
41. James, Clive (1996). "Global Review of the Field Testing and Commercialization of Transgenic Plants: 1986 to 1995" (PDF). The International Service for the Acquisition of Agri-biotech Applications. Retrieved 17 July 2010.
42. James, Clive (1997). "Global Status of Transgenic Crops in 1997"(PDF). ISAAA Briefs No. 5.: 31.
43. Bruening, G.; Lyons, J.M. (2000). "The case of the FLAVR SAVR tomato". California Agriculture. 54 (4): 6–7. doi:10.3733/ca.v054n04p6.
44. MacKenzie, Debora (18 June 1994). "Transgenic tobacco is European first". New Scientist.
45. Genetically Altered Potato Ok'd For Crops Lawrence Journal-World - 6 May 1995
46. Global Status of Commercialized Biotech/GM Crops: 2009 ISAAA Brief 41-2009, 23 February 2010. Retrieved 10 August 2010
47. Pennisi, Elizabeth (2010-05-21). "Synthetic Genome Brings New Life to Bacterium". Science. 328 (5981): 958–959. doi:10.1126/science.328.5981.958. ISSN 0036-8075. PMID 20488994.
48. Gibson, D. G.; Glass, J. I.; Lartigue, C.; Noskov, V. N.; Chuang, R.-Y.; Algire, M. A.; Benders, G. A.; Montague, M. G.; Ma, L.; Moodie, M. M.; Merryman, C.; Vashee, S.; Krishnakumar, R.; Assad-Garcia, N.; Andrews-Pfannkoch, C.; Denisova, E. A.; Young, L.; Qi, Z.-Q.; Segall-Shapiro, T. H.; Calvey, C. H.; Parmar, P. P.; Hutchison Ca, C. A.; Smith, H. O.; Venter, J. C. (2010). "Creation of a Bacterial Cell Controlled by a Chemically Synthesized Genome". Science. 329 (5987): 52–6. doi:10.1126/science.1190719. PMID 20488990.
49. Malyshev, Denis A.; Dhami, Kirandeep; Lavergne, Thomas; Chen, Tingjian; Dai,

Nan; Foster, Jeremy M.; Corrêa, Ivan R.; Romesberg, Floyd E. (2014-05-15). "A semi-synthetic organism with an expanded genetic alphabet". Nature. 509 (7500): 385–388. doi:10.1038/nature13314. ISSN 0028-0836. PMC 4058825 . PMID 24805238.

50. Thyer, Ross; Ellefson, Jared (2014-05-15). "Synthetic biology: New letters for life's alphabet". Nature. 509 (7500): 291–292. doi:10.1038/nature13335. ISSN 0028-0836.
51. Alberts B, Johnson A, Lewis J, et al. (2002). "8". Isolating, Cloning, and Sequencing DNA. (4th ed.). New York: Garland Science.
52. Kaufman, R I; Nixon, B T (1996). "Use of PCR to isolate genes encoding sigma54-dependent activators from diverse bacteria". J Bacteriol. 178 (13): 3967–3970. doi:10.1128/jb.178.13.3967-3970.1996. PMC 232662 . PMID 8682806.
53. Liang, Jing; Luo, Yunzi; Zhao, Huimin (2011). "Synthetic biology: Putting synthesis into biology". Wiley Interdisciplinary Reviews: Systems Biology and Medicine. 3: 7–20. doi:10.1002/wsbm.104.
54. Berg, P.; Mertz, J. E. (2010). "Personal Reflections on the Origins and Emergence of Recombinant DNA Technology". Genetics. 184 (1): 9–17. doi:10.1534/genetics.109.112144. PMC 2815933 . PMID 20061565.
55. Townsend JA, Wright DA, Winfrey RJ, et al. (May 2009). "High-frequency modification of plant genes using engineered zinc-finger nucleases". Nature. 459 (7245): 442–5. Bibcode:2009Natur.459..442T. doi:10.1038/nature07845. PMC 2743854 . PMID 19404258.
56. Shukla VK, Doyon Y, Miller JC, et al. (May 2009). "Precise genome modification in the crop species Zea mays using zinc-finger nucleases". Nature. 459 (7245): 437–41. Bibcode:2009Natur.459..437S. doi:10.1038/nature07992. PMID 19404259.
57. Grizot S, Smith J, Daboussi F, et al. (September 2009). "Efficient targeting of a SCID gene by an engineered single-chain homing endonuclease". Nucleic Acids Res. 37 (16): 5405–19. doi:10.1093/nar/gkp548. PMC 2760784 . PMID 19584299.
58. Gao H, Smith J, Yang M, et al. (January 2010). "Heritable targeted mutagenesis in maize using a designed endonuclease". Plant J. 61 (1): 176–87. doi:10.1111/j.1365-313X.2009.04041.x. PMID 19811621.
59. Christian M, Cermak T, Doyle EL, et al. (July 2010). "TAL Effector Nucleases Create Targeted DNA Double-strand Breaks". Genetics. 186 (2): 757–61. doi:10.1534/genetics.110.120717. PMC 2942870 . PMID 20660643.
60. Li T, Huang S, Jiang WZ, et al. (August 2010). "TAL nucleases (TALNs): hybrid proteins composed of TAL effectors and FokI DNA-cleavage domain". Nucleic Acids Res. 39 (1): 359–72. doi:10.1093/nar/gkq704. PMC 3017587 . PMID 20699274.
61. Ekker, S.C. (2008). "Zinc finger-based knockout punches for zebrafish genes". Zebrafish. 5 (2): 1121–3. doi:10.1089/zeb.2008.9988. PMC 2849655 . PMID 18554175.

62. Geurts AM, Cost GJ, Freyvert Y, et al. (July 2009). "Knockout rats via embryo microinjection of zinc-finger nucleases". Science. 325 (5939): 433. Bibcode:2009Sci...325..433G. doi:10.1126/science.1172447. PMC 2831805 . PMID 19628861.
63. Chen, I; Dubnau, D (2004). "DNA uptake during bacterial transformation". Nat. Rev. Microbiol. 2 (3): 241–9. doi:10.1038/nrmicro844. PMID 15083159.
64. Head, Graham; Hull, Roger H; Tzotzos, George T. (2009). Genetically Modified Plants: Assessing Safety and Managing Risk. London: Academic Pr. p. 244. ISBN 0-12-374106-8.
65. Gelvin, S. B. (2003). "Agrobacterium-Mediated Plant Transformation: The Biology behind the "Gene-Jockeying" Tool". Microbiology and Molecular Biology Reviews. 67 (1): 16–37, table of contents. doi:10.1128/MMBR.67.1.16-37.2003. PMC 150518 . PMID 12626681.
66. Darbani, Behrooz; Farajnia, Safar; Toorchi, Mahmoud; Zakerbostanabad, Saeed; Noeparvar, Shahin; Stewart, Jr., C. Neal (2010). "DNA-Delivery Methods to Produce Transgenic Plants". Science Alert.
67. Hohn, Barbara; Levy, Avraham A; Puchta, Holger (2001). "Elimination of selection markers from transgenic plants". Current Opinion in Biotechnology. 12 (2): 139–43. doi:10.1016/S0958-1669(00)00188-9. PMID 11287227.
68. Esvelt, KM.; Wang, HH. (2013). "Genome-scale engineering for systems and synthetic biology". Mol Syst Biol. 9: 641. doi:10.1038/msb.2012.66. PMC 3564264 . PMID 23340847.
69. Tan, WS.; Carlson, DF.; Walton, MW.; Fahrenkrug, SC.; Hackett, PB. (2012). "Precision editing of large animal genomes". Adv Genet. Advances in Genetics. 80: 37–97. doi:10.1016/B978-0-12-404742-6.00002-8. ISBN 9780124047426. PMC 3683964 . PMID 23084873.
70. "Natural Genetic Engineering and Natural Genome Editing". Annals of the New York Academy of Sciences. 1178. 2009.
71. Avise, John C. (2004). The hope, hype & reality of genetic engineering: remarkable stories from agriculture, industry, medicine, and the environment. Oxford University Press US. p. 22. ISBN 978-0-19-516950-8.
72. "Engineering algae to make complex anti-cancer 'designer' drug". PhysOrg. 10 December 2012. Retrieved 15 April 2013.
73. =Roque, AC; Lowe, CR; Taipa, MA. (2004). "Antibodies and genetically engineered related molecules: production and purification". Biotechnol Proress. 20 (3): 639–54. doi:10.1021/bp030070k. PMID 15176864.
74. Rodriguez, Luis L.; Grubman, Marvin J. (2009). "Foot and mouth disease virus vaccines". Vaccine. 27: D90–4. doi:10.1016/j.vaccine.2009.08.039. PMID 19837296.
75. "Background: Cloned and Genetically Modified Animals". Center for Genetics and Society. 14 April 2005.
76. "Knockout Mice". Nation Human Genome Research Institute. 2009.

77. "GM pigs best bet for organ transplant". Medical News Today. 21 September 2003.

78. Fischer, Alain; Hacein-Bey-Abina, Salima; Cavazzana-Calvo, Marina (2010). "20 years of gene therapy for SCID". Nature Immunology. 11 (6): 457–60. doi:10.1038/ni0610-457. PMID 20485269.

79. Ledford, Heidi (2011). "Cell therapy fights leukaemia". Nature. doi:10.1038/news.2011.472.

80. Lewitt, Peter A; Rezai, Ali R; Leehey, Maureen A; Ojemann, Steven G; Flaherty, Alice W; Eskandar, Emad N; Kostyk, Sandra K; Thomas, Karen; Sarkar, Atom; Siddiqui, Mustafa S; Tatter, Stephen B; Schwalb, Jason M; Poston, Kathleen L; Henderson, Jaimie M; Kurlan, Roger M; Richard, Irene H; Van Meter, Lori; Sapan, Christine V; During, Matthew J; Kaplitt, Michael G; Feigin, Andrew (2011). "AAV2-GAD gene therapy for advanced Parkinson's disease: A double-blind, sham-surgery controlled, randomised trial". The Lancet Neurology. 10 (4): 309–19. doi:10.1016/S1474-4422(11)70039-4. PMID 21419704.

81. Gallagher, James. (2 November 2012) BBC News – Gene therapy: Glybera approved by European Commission. Bbc.co.uk. Retrieved on 15 December 2012.

82. Richards, Sabrina. "Gene Therapy Arrives in Europe". The Scientist. Retrieved 16 November 2012.

83. "1990 The Declaration of Inuyama". 5 August 2001. Archived from the original on 5 August 2001.

84. Smith KR, Chan S, Harris J. Human germline genetic modification: scientific and bioethical perspectives. Arch Med Res. 2012 Oct;43(7):491-513. doi: 10.1016/j.arcmed.2012.09.003. PMID 23072719

85. Wade, Nicholas (19 March 2015). "Scientists Seek Ban on Method of Editing the Human Genome". New York Times. Retrieved 20 March 2015.

86. Pollack, Andrew (3 March 2015). "A Powerful New Way to Edit DNA". New York Times. Retrieved 20 March 2015.

87. Baltimore, David; Berg, Paul; Botchan, Dana; Charo, R. Alta; Church, George; Corn, Jacob E.; Daley, George Q.; Doudna, Jennifer A.; Fenner, Marsha; Greely, Henry T.; Jinek, Martin; Martin, G. Steven; Penhoet, Edward; Puck, Jennifer; Sternberg, Samuel H.; Weissman, Jonathan S.; Yamamoto, Keith R. (19 March 2015). "A prudent path forward for genomic engineering and germline gene modification". Science. 348: 36–8. doi:10.1126/science.aab1028. PMC 4394183 . PMID 25791083. Retrieved 20 March 2015.

88. Lanphier, Edward; Urnov, Fyodor; Haecker, Sarah Ehlen; Werner, Michael; Smolenski, Joanna (26 March 2015). "Don't edit the human germ line". Nature. 519: 410–411. doi:10.1038/519410a. PMID 25810189. Retrieved 20 March 2015.

89. Kolata, Gina (23 April 2015). "Chinese Scientists Edit Genes of Human Embryos, Raising Concerns". New York Times. Retrieved 24 April 2015.

90. Liang, Puping; et al. (18 April 2015). "CRISPR/Cas9-mediated gene editing in human tripronuclear zygotes". Protein & Cell. 6: 363–72. doi:10.1007/s13238-

015-0153-5. PMC 4417674 . PMID 25894090. Retrieved 24 April 2015.

91. Wade, Nicholas (3 December 2015). "Scientists Place Moratorium on Edits to Human Genome That Could Be Inherited". New York Times. Retrieved 3 December 2015.
92. Bergeson, Emilie R. (1997). "The Ethics of Gene Therapy".
93. Hanna, Kathi E. "Genetic Enhancement". National Human Genome Research Institute.
94. "Applications of Genetic Engineering". Microbiologyprocedure. Retrieved 9 July 2010.
95. "Biotech: What are transgenic organisms?". Easyscience. 2002. Retrieved 9 July 2010.
96. Savage, Neil (1 August 2007). "Making Gasoline from Bacteria: A biotech startup wants to coax fuels from engineered microbes". Technology Review. Retrieved 16 July 2015.
97. Summers, Rebecca (24 April 2013) Bacteria churn out first ever petrol-like biofuel New Scientist, Retrieved 27 April 2013
98. "Applications of Some Genetically Engineered Bacteria". Retrieved 9 July 2010.
99. Sanderson, Katherine (24 February 2012) New Portable Kit Detects Arsenic In Wells Chemical and Engineering News, Retrieved 23 January 2013
100. Reece, Jane B.; Urry, Lisa A.; Cain, Michael L.; Wasserman, Steven A.; Minorsky, Peter V.; Jackson, Robert B. (2011). Campbell Biology Ninth Edition. San Francisco: Pearson Benjamin Cummings. p. 421. ISBN 0-321-55823-5.
101. "New virus-built battery could power cars, electronic devices". Web.mit.edu. 2 April 2009. Retrieved 17 July 2010.
102. "Hidden Ingredient In New, Greener Battery: A Virus". Npr.org. Retrieved 17 July 2010.
103. "Researchers Synchronize Blinking 'Genetic Clocks' -- Genetically Engineered Bacteria That Keep Track of Time". ScienceDaily. 24 January 2010.
104. Suszkiw, Jan (November 1999). "Tifton, Georgia: A Peanut Pest Showdown". Agricultural Research magazine. Retrieved 23 November 2008.
105. Magaña-Gómez, JA; de la Barca, A.M. (2009). "Risk assessment of genetically modified crops for nutrition and health". Nutr. Rev. 67 (1): 1–16. doi:10.1111/j.1753-4887.2008.00130.x. PMID 19146501.
106. Islam, Aparna (2008). "Fungus Resistant Transgenic Plants: Strategies, Progress and Lessons Learnt". Plant Tissue Culture and Biotechnology. 16 (2): 117–38. doi:10.3329/ptcb.v16i2.1113.
107. "Disease resistant crops". GMO Compass.
108. Demont, M; Tollens, E (2004). "First impact of biotechnology in the EU: Bt maize adoption in Spain". Annals of Applied Biology. 145 (2): 197–207. doi:10.1111/j.1744-7348.2004.tb00376.x.
109. Whitman, Deborah B. (2000). "Genetically Modified Foods: Harmful or Helpful?".

110. Young, Emma (2003). "GM cows to please cheese-makers". New Scientist.

111. Rapeseed (canola) has been genetically engineered to modify its oil content with a gene encoding a "12:0 thioesterase" (TE) enzyme from the California bay plant (Umbellularia californica) to increase medium length fatty acids, see: Geo-pie.cornell.edu

112. Bomgardner Melody M (2012). "Replacing Trans Fat: New crops from Dow Chemical and DuPont target food makers looking for stable, heart-healthy oils". Chemical and Engineering News. 90 (11): 30–32.

113. Marvier, Michelle (2008). "Pharmaceutical crops in California, benefits and risks. A review". Agronomy for Sustainable Development. 28 (1): 1–9. doi:10.1051/agro:2007050.

114. "FDA Approves First Human Biologic Produced by GE Animals". US Food and Drug Administration.

115. Rebêlo, Paulo (15 July 2004). "GM cow milk 'could provide treatment for blood disease'". SciDev.

116. Pollack, Andrew (19 November 2015). "Genetically Engineered Salmon Approved for Consumption". The New York Times. Retrieved 21 April 2016.

117. "Giant GM salmon on the way". BBC News. 11 April 2000.

118. Chivian, Eric; Bernstein, Aaron (2008). Sustaining Life. Oxford University Press, Inc. ISBN 978-0-19-517509-7.

119. Carrington, Damien (13 June 2012) GM crops good for environment, study finds The Guardian. Retrieved 16 June 2012

120. Nicolia, Alessandro; Manzo, Alberto; Veronesi, Fabio; Rosellini, Daniele (2013). "An overview of the last 10 years of genetically engineered crop safety research" (PDF). Critical Reviews in Biotechnology. 34: 1–12. doi:10.3109/07388551.2013.823595. PMID 24041244. We have reviewed the scientific literature on GE crop safety for the last 10 years that catches the scientific consensus matured since GE plants became widely cultivated worldwide, and we can conclude that the scientific research conducted so far has not detected any significant hazard directly connected with the use of GM crops.

121. "State of Food and Agriculture 2003–2004. Agricultural Biotechnology: Meeting the Needs of the Poor. Health and environmental impacts of transgenic crops". Food and Agriculture Organization of the United Nations. Retrieved February 8, 2016. Currently available transgenic crops and foods derived from them have been judged safe to eat and the methods used to test their safety have been deemed appropriate. These conclusions represent the consensus of the scientific evidence surveyed by the ICSU (2003) and they are consistent with the views of the World Health Organization (WHO, 2002). These foods have been assessed for increased risks to human health by several national regulatory authorities (inter alia, Argentina, Brazil, Canada, China, the United Kingdom and the United States) using their national food safety procedures (ICSU). To date no verifiable untoward toxic or nutritionally deleterious effects resulting from the consumption of foods derived from genetically modified crops have

been discovered anywhere in the world (GM Science Review Panel). Many millions of people have consumed foods derived from GM plants - mainly maize, soybean and oilseed rape - without any observed adverse effects (ICSU).

122. Ronald, Pamela (May 5, 2011). "Plant Genetics, Sustainable Agriculture and Global Food Security". Genetics. 188: 11–20. doi:10.1534/genetics.111.128553. PMC 3120150 . PMID 21546547. There is broad scientific consensus that genetically engineered crops currently on the market are safe to eat. After 14 years of cultivation and a cumulative total of 2 billion acres planted, no adverse health or environmental effects have resulted from commercialization of genetically engineered crops (Board on Agriculture and Natural Resources, Committee on Environmental Impacts Associated with Commercialization of Transgenic Plants, National Research Council and Division on Earth and Life Studies 2002). Both the U.S. National Research Council and the Joint Research Centre (the European Union's scientific and technical research laboratory and an integral part of the European Commission) have concluded that there is a comprehensive body of knowledge that adequately addresses the food safety issue of genetically engineered crops (Committee on Identifying and Assessing Unintended Effects of Genetically Engineered Foods on Human Health and National Research Council 2004; European Commission Joint Research Centre 2008). These and other recent reports conclude that the processes of genetic engineering and conventional breeding are no different in terms of unintended consequences to human health and the environment (European Commission Directorate-General for Research and Innovation 2010).

123. "A decade of EU-funded GMO research (2001–2010)" (PDF). Directorate-General for Research and Innovation. Biotechnologies, Agriculture, Food. European Commission, European Union. 2010. doi:10.2777/97784. ISBN 978-92-79-16344-9. Retrieved February 8, 2016.

124. "AMA Report on Genetically Modified Crops and Foods (online summary)". American Medical Association. January 2001. Retrieved March 19, 2016. A report issued by the scientific council of the American Medical Association (AMA) says that no long-term health effects have been detected from the use of transgenic crops and genetically modified foods, and that these foods are substantially equivalent to their conventional counterparts. (from online summary prepared by ISAAA)" "Crops and foods produced using recombinant DNA techniques have been available for fewer than 10 years and no long-term effects have been detected to date. These foods are substantially equivalent to their conventional counterparts. (from original report by AMA:)

125. "REPORT 2 OF THE COUNCIL ON SCIENCE AND PUBLIC HEALTH (A-12): Labeling of Bioengineered Foods" (PDF). American Medical Association. 2012. Archived from the original on 7 September 2012. Retrieved March 19, 2016. Bioengineered foods have been consumed for close to 20 years, and during that time, no overt consequences on human health have been reported and/or substantiated in the peer-reviewed literature.

126. "Restrictions on Genetically Modified Organisms: United States. Public and Scholarly Opinion". Library of Congress. June 9, 2015. Retrieved February

8, 2016. Several scientific organizations in the US have issued studies or statements regarding the safety of GMOs indicating that there is no evidence that GMOs present unique safety risks compared to conventionally bred products. These include the National Research Council, the American Association for the Advancement of Science, and the American Medical Association. Groups in the US opposed to GMOs include some environmental organizations, organic farming organizations, and consumer organizations. A substantial number of legal academics have criticized the US's approach to regulating GMOs.

127. "Genetically Engineered Crops: Experiences and Prospects". The National Academies of Sciences, Engineering, and Medicine (US). 2016. p. 149. Retrieved May 19, 2016. Overall finding on purported adverse effects on human health of foods derived from GE crops: On the basis of detailed examination of comparisons of currently commercialized GE with non-GE foods in compositional analysis, acute and chronic animal toxicity tests, long-term data on health of livestock fed GE foods, and human epidemiological data, the committee found no differences that implicate a higher risk to human health from GE foods than from their non-GE counterparts.

128. "Frequently asked questions on genetically modified foods". World Health Organization. Retrieved February 8, 2016. Different GM organisms include different genes inserted in different ways. This means that individual GM foods and their safety should be assessed on a case-by-case basis and that it is not possible to make general statements on the safety of all GM foods.

129. GM foods currently available on the international market have passed safety assessments and are not likely to present risks for human health. In addition, no effects on human health have been shown as a result of the consumption of such foods by the general population in the countries where they have been approved. Continuous application of safety assessments based on the Codex Alimentarius principles and, where appropriate, adequate post market monitoring, should form the basis for ensuring the safety of GM foods.

130. Haslberger, Alexander G. (2003). "Codex guidelines for GM foods include the analysis of unintended effects". Nature Biotechnolgy. 21: 739–741. doi:10.1038/nbt0703-739. PMID 12833088. These principles dictate a case-by-case premarket assessment that includes an evaluation of both direct and unintended effects.

131. Some medical organizations, including the British Medical Association, advocate further caution based upon the precautionary principle:

132. "Genetically modified foods and health: a second interim statement"(PDF). British Medical Association. March 2004. Retrieved March 21, 2016. In our view, the potential for GM foods to cause harmful health effects is very small and many of the concerns expressed apply with equal vigour to conventionally derived foods. However, safety concerns cannot, as yet, be dismissed completely on the basis of information currently available.

133. When seeking to optimise the balance between benefits and risks, it is prudent to err on the side of caution and, above all, learn from accumulating knowledge and experience. Any new technology such as genetic modification must be

examined for possible benefits and risks to human health and the environment. As with all novel foods, safety assessments in relation to GM foods must be made on a case-by-case basis.

134. Members of the GM jury project were briefed on various aspects of genetic modification by a diverse group of acknowledged experts in the relevant subjects. The GM jury reached the conclusion that the sale of GM foods currently available should be halted and the moratorium on commercial growth of GM crops should be continued. These conclusions were based on the precautionary principle and lack of evidence of any benefit. The Jury expressed concern over the impact of GM crops on farming, the environment, food safety and other potential health effects.
135. The Royal Society review (2002) concluded that the risks to human health associated with the use of specific viral DNA sequences in GM plants are negligible, and while calling for caution in the introduction of potential allergens into food crops, stressed the absence of evidence that commercially available GM foods cause clinical allergic manifestations. The BMA shares the view that that there is no robust evidence to prove that GM foods are unsafe but we endorse the call for further research and surveillance to provide convincing evidence of safety and benefit.
136. Funk, Cary; Rainie, Lee (January 29, 2015). "Public and Scientists' Views on Science and Society". Pew Research Center. Retrieved February 24, 2016. The largest differences between the public and the AAAS scientists are found in beliefs about the safety of eating genetically modified (GM) foods. Nearly nine-in-ten (88%) scientists say it is generally safe to eat GM foods compared with 37% of the general public, a difference of 51 percentage points.
137. Marris, Claire (2001). "Public views on GMOs: deconstructing the myths" (PDF). EMBO Reports. 2: 545–548. doi:10.1093/embo-reports/kve142. PMC 1083956 . PMID 11463731.
138. Final Report of the PABE research project (December 2001). "Public Perceptions of Agricultural Biotechnologies in Europe". Commission of European Communities. Retrieved February 24, 2016.
139. Scott, Sydney E.; Inbar, Yoel; Rozin, Paul (2016). "Evidence for Absolute Moral Opposition to Genetically Modified Food in the United States" (PDF). Perspectives on Psychological Science. 11 (3): 315–324. doi:10.1177/1745691615621275. PMID 27217243.
140. "Restrictions on Genetically Modified Organisms". Library of Congress. June 9, 2015. Retrieved February 24, 2016.
141. Bashshur, Ramona (February 2013). "FDA and Regulation of GMOs". American Bar Association. Retrieved February 24, 2016.
142. Sifferlin, Alexandra (October 3, 2015). "Over Half of E.U. Countries Are Opting Out of GMOs". Time.
143. Lynch, Diahanna; Vogel, David (April 5, 2001). "The Regulation of GMOs in Europe and the United States: A Case-Study of Contemporary European

Regulatory Politics". Council on Foreign Relations. Retrieved February 24, 2016.

144. "Can GM crops harm the environment?". National Environment Research Council (NERC).
145. Angulo, E.; Cooke, B. (2002). "First synthesize new viruses then regulate their release? The case of the wild rabbit". Molecular Ecology. 11: 2703–9. doi:10.1046/j.1365-294X.2002.01635.x. PMID 12453252.
146. Adams; et al. (2 August 2002). "The Case for Genetic Engineering of Native and Landscape Trees against Introduced Pests and Diseases". Conservation Biology. 16: 874–879. doi:10.1046/j.1523-1739.2002.00523.x. Retrieved 16 May 2016.
147. Thomas; et al. (25 September 2013). "Ecology: Gene tweaking for conservation". Nature. 501: 485–6. doi:10.1038/501485a. PMID 24073449. Retrieved 16 May 2016.
148. Pasko, Jessica M. (2007-03-04). "Bio-artists bridge gap between arts, sciences: Use of living organisms is attracting attention and controversy". msnbc.
149. Jackson, Joab (6 December 2005). "Genetically Modified Bacteria Produce Living Photographs". National Geographic News.
150. Phys.Org website. 4 April 2005 "Plant gene replacement results in the world's only blue rose".
151. Katsumoto, Yukihisa; Fukuchi-Mizutani, Masako; Fukui, Yuko; Brugliera, Filippa; Holton, Timothy A.; Karan, Mirko; Nakamura, Noriko; Yonekura-Sakakibara, Keiko; Togami, Junichi; Pigeaire, Alix; Tao, Guo-Qing; Nehra, Narender S.; Lu, Chin-Yi; Dyson, Barry K.; Tsuda, Shinzo; Ashikari, Toshihiko; Kusumi, Takaaki; Mason, John G.; Tanaka, Yoshikazu (2007). "Engineering of the Rose Flavonoid Biosynthetic Pathway Successfully Generated Blue-Hued Flowers Accumulating Delphinidin". Plant and Cell Physiology. 48 (11): 1589–600. doi:10.1093/pcp/pcm131. PMID 17925311.
152. Published PCT Application WO2000049150 "Chimeric Gene Constructs for Generation of Fluorescent Transgenic Ornamental Fish." National University of Singapore
153. Stewart, C. Neal (2006). "Go with the glow: Fluorescent proteins to light transgenic organisms". Trends in Biotechnology. 24 (4): 155–62. doi:10.1016/j.tibtech.2006.02.002. PMID 16488034.
154. WHO (1987): Principles for the Safety Assessment of Food Additives and Contaminants in Food, Environmental Health Criteria 70. World Health Organization, Geneva
155. WHO (1991): Strategies for assessing the safety of foods produced by biotechnology, Report of a Joint FAO/WHO Consultation. World Health Organization, Geneva ISBN 9789241561457. PDF download library.health.go.ug/download/file/fid/790 here]
156. WHO (1993): Health aspects of marker genes in genetically modified plants, Report of a WHO Workshop. World Health Organization, Geneva
157. WHO (1995): Application of the principle of substantial equivalence to the

safety evaluation of foods or food components from plants derived by modern biotechnology, Report of a WHO Workshop. World Health Organization, Geneva

158. Skoglund, Pontus; Ersmark, Erik; Palkopoulou, Eleftheria; Dalén, Love (2015-06-01). "Ancient Wolf Genome Reveals an Early Divergence of Domestic Dog Ancestors and Admixture into High-Latitude Breeds". Current Biology. 25 (11): 1515–1519. doi:10.1016/j.cub.2015.04.019. PMID 26004765.
159. Jackson, DA; Symons, RH; Berg, P (1 October 1972). "Biochemical Method for Inserting New Genetic Information into DNA of Simian Virus 40: Circular SV40 DNA Molecules Containing Lambda Phage Genes and the Galactose Operon of Escherichia coli". PNAS. 69 (10): 2904–2909. Bibcode:1972PNAS...69.2904J. doi:10.1073/pnas.69.10.2904. PMC 389671 . PMID 4342968.
160. Larson, Greger; Karlsson, Elinor K.; Perri, Angela; Webster, Matthew T.; Ho, Simon Y. W.; Peters, Joris; Stahl, Peter W.; Piper, Philip J.; Lingaas, Frode (2012-06-05). "Rethinking dog domestication by integrating genetics, archeology, and biogeography". Proceedings of the National Academy of Sciences. 109 (23): 8878–8883. doi:10.1073/pnas.1203005109. PMC 3384140 . PMID 22615366.
161. Montague, Michael J.; Li, Gang; Gandolfi, Barbara; Khan, Razib; Aken, Bronwen L.; Searle, Steven M. J.; Minx, Patrick; Hillier, LaDeana W.; Koboldt, Daniel C. (2014-12-02). "Comparative analysis of the domestic cat genome reveals genetic signatures underlying feline biology and domestication". Proceedings of the National Academy of Sciences. 111 (48): 17230–17235. doi:10.1073/pnas.1410083111. PMC 4260561 . PMID 25385592.
162. Zeder, Melinda A. (2008-08-19). "Domestication and early agriculture in the Mediterranean Basin: Origins, diffusion, and impact". Proceedings of the National Academy of Sciences. 105 (33): 11597–11604. doi:10.1073/pnas.0801317105. PMC 2575338 . PMID 18697943.
163. Zohary, & Hopf Weiss, p. 1.
164. the history of maize cultivation in southern Mexico dates back 9000 years. New York Times, (2010-05-25)
165. Colledge, Sue; Conolly, James (2007). The Origins and Spread of Domestic Plants in Southwest Asia and Europe. p. 40. ISBN 1598749889.
166. Zohary, & Hopf Weiss, p. 5.
167. Zohary, & Hopf Weiss, p. 6.
168. Kingsbury, Noel (2009). Hybrid: The History and Science of Plant Breeding. University of Chicago Press. ISBN 0226437051.
169. "Evolution of Wheatpublisher=Wheat, the big picture".
170. Le Page, Michael (2016-03-17). "Farmers may have been accidentally making GMOs for millennia". The New Scientist. Retrieved 2016-07-11.
171. Schouten, H. J.; Jacobsen, E. (2007). "Are Mutations in Genetically Modified Plants Dangerous?". Journal of Biomedicine and Biotechnology. 2007: 1–2. doi:10.1155/2007/82612.
172. Hartl, D. L.; Orel, V. (1992). "What Did Gregor Mendel Think He Discovered?".

Genetics. 131 (2): 245–25. PMC 1205000 . PMID 1644269.

173. Vries, H. de (1889) Intracellular Pangenesis ("pan-gene" definition on page 7 and 40 of this 1910 translation in English)
174. Creative Sponge. "The Bateson Lecture".
175. Roberts, R. J. (2005). "Classic Perspective: How restriction enzymes became the workhorses of molecular biology". Proceedings of the National Academy of Sciences. 102 (17): 5905–5908. doi:10.1073/pnas.0500923102. PMC 1087929 . PMID 15840723.
176. Weiss, B.; Richardson, C. C. (1967). "Enzymatic breakage and joining of deoxyribonucleic acid, I. Repair of single-strand breaks in DNA by an enzyme system from Escherichia coli infected with T4 bacteriophage". Proceedings of the National Academy of Sciences. 57 (4): 1021–8. doi:10.1073/pnas.57.4.1021. PMC 224649 . PMID 5340583.
177. Lederberg, J (1952). "Cell genetics and hereditary symbiosis". Physiological reviews. 32 (4): 403–30. PMID 13003535.
178. Mandel, Morton; Higa, Akiko (1970). "Calcium-dependent bacteriophage DNA infection". Journal of Molecular Biology. 53 (1): 159–162. doi:10.1016/0022-2836(70)90051-3. PMID 4922220.
179. Cohen, S. N.; Chang, A. C. Y.; Hsu, L. (1972). "Non chromosomal Antibiotic Resistance in Bacteria: Genetic Transformation of Escherichia coli by R-Factor DNA". Proceedings of the National Academy of Sciences. 69 (8): 2110–4. doi:10.1073/pnas.69.8.2110. PMC 426879 . PMID 4559594.
180. Wirth, Reinhard; Friesenegger, Anita; Fiedlerand, Stefan (1989). "Transformation of various species of gram-negative bacteria belonging to 11 different genera by electroporation". Molecular and General Genetics MGG. 216 (1): 175–177. doi:10.1007/BF00332248. PMID 2659971.
181. Nester, Eugene (2008). "Agrobacterium: The Natural Genetic Engineer (100 Years Later)".
182. Zambryski, P.; Joos, H.; Genetello, C.; Leemans, J.; Montagu, M. V.; Schell, J. (1983). "Ti plasmid vector for the introduction of DNA into plant cells without alteration of their normal regeneration capacity". The EMBO Journal. 2 (12): 2143–2150. PMC 555426 . PMID 16453482.
183. Jackson, D. A.; Symons, R. H.; Berg, P. (1972). "Biochemical Method for Inserting New Genetic Information into DNA of Simian Virus 40: Circular SV40 DNA Molecules Containing Lambda Phage Genes and the Galactose Operon of Escherichia coli". Proceedings of the National Academy of Sciences. 69 (10): 2904–9. doi:10.1073/pnas.69.10.2904. PMC 389671 . PMID 4342968.
184. "Genome and genetics timeline – 1973". Genome news network.
185. Arnold, Paul (2009). "History of Genetics: Genetic Engineering Timeline".
186. Cohen, Stanley N.; Chang, Annie C. Y. (1973). "Recircularization and Autonomous Replication of a Sheared R-Factor DNA Segment inEscherichia coli Transformants". Proceedings of the National Academy of Sciences of the

United States of America. 70 (5): 1293–1297. doi:10.1073/pnas.70.5.1293. JSTOR 62105. PMC 433482 . PMID 4576014.

187. Jaenisch, R.; Mintz, B. (1974). "Simian virus 40 DNA sequences in DNA of healthy adult mice derived from preimplantation blastocysts injected with viral DNA". Proceedings of the National Academy of Sciences of the United States of America. 71 (4): 1250–4. doi:10.1073/pnas.71.4.1250. PMC 388203 . PMID 4364530.
188. "'Any idiot can do it.' Genome editor CRISPR could put mutant mice in everyone's reach". Science | AAAS. 2016-11-02. Retrieved 2016-12-02.
189. Brownlee, C. (2004). "Inaugural Article: Biography of Rudolf Jaenisch". Proceedings of the National Academy of Sciences. 101 (39): 13982–13984. doi:10.1073/pnas.0406416101. PMC 521108 . PMID 15383657.
190. Gordon, J.; Ruddle, F. (1981). "Integration and stable germ line transmission of genes injected into mouse pronuclei". Science. 214 (4526): 1244–6. Bibcode:1981Sci...214.1244G. doi:10.1126/science.6272397. PMID 6272397.
191. Costantini, F.; Lacy, E. (1981). "Introduction of a rabbit β-globin gene into the mouse germ line". Nature. 294 (5836): 92–4. Bibcode:1981Natur.294...92C. doi:10.1038/294092a0. PMID 6945481.
192. Lemaux, P. (2008). "Genetically Engineered Plants and Foods: A Scientist's Analysis of the Issues (Part I)". Annual Review of Plant Biology. 59: 771–812. doi:10.1146/annurev.arplant.58.032806.103840. PMID 18284373.
193. Bevan, M. W.; Flavell, R. B.; Chilton, M. D. (1983). "A chimaeric antibiotic resistance gene as a selectable marker for plant cell transformation". Nature. 304 (5922): 184–187. Bibcode:1983Natur.304..184B. doi:10.1038/304184a0.
194. Berg, P.; Baltimore, D.; Brenner, S.; Roblin, R. O.; Singer, M. F. (1975). "Summary statement of the Asilomar conference on recombinant DNA molecules". Proceedings of the National Academy of Sciences. 72 (6): 1981–4. doi:10.1073/pnas.72.6.1981. PMC 432675 . PMID 806076.
195. Hutt, P. B. (1978). "Research on recombinant DNA molecules: The regulatory issues". Southern California law review. 51 (6): 1435–50. PMID 11661661.
196. McHughen A, Smyth S (2008). "US regulatory system for genetically modified genetically modified organism (GMO), rDNA or transgenic crop cultivars". Plant biotechnology journal. 6 (1): 2–12. doi:10.1111/j.1467-7652.2007.00300.x. PMID 17956539.
197. Bull, A.T., Holt, G. and Lilly, M.D. (1982). Biotechnology : international trends and perspectives (PDF). Paris: Organisation for Economic Co-operation and Development.
198. U.S. Office of Science and Technology Policy (1986). "Coordinated framework for regulation of biotechnology; announcement of policy; notice for public comment". Federal register. 51 (123): 23302–50. PMID 11655807.
199. WHO (1987) Principles for the Safety Assessment of Food Additives and Contaminants in Food, Environmental Health Criteria 70. World Health

Organization, Geneva

200. WHO (1991) Strategies for assessing the safety of foods produced by biotechnology, Report of a Joint FAO/WHO Consultation. World Health Organization, Geneva

201. WHO (1993) Health aspects of marker genes in genetically modified plants, Report of a WHO Workshop. World Health Organization, Geneva

202. WHO (1995) Application of the principle of substantial equivalence to the safety evaluation of foods or food components from plants derived by modern biotechnology, Report of a WHO Workshop. World Health Organization, Geneva

203. Gruère, Colin A. Carter and Guillaume P. (2003-12-15). "Mandatory Labeling of Genetically Modified Foods: Does it Really Provide Consumer Choice?". www.agbioforum.org. Retrieved 2016-01-21.

204. Strom, Stephanie (2013-06-03). "Connecticut Approves Qualified Genetic Labeling". The New York Times. Retrieved 2016-01-21.

205. "Knockout Mice". National Human Genome Research Institute.

206. Hanahan, D.; Wagner, E. F.; Palmiter, R. D. (2007). "The origins of oncomice: A history of the first transgenic mice genetically engineered to develop cancer". Genes & Development. 21 (18): 2258–2270. doi:10.1101/gad.1583307. PMID 17875663.

207. Pilcher, Helen R. (2003). "It's a knockout: First rat to have key genes altered". Nature. doi:10.1038/news030512-17.

208. Zan, Y; Haag, J. D.; Chen, K. S.; Shepel, L. A.; Wigington, D; Wang, Y. R.; Hu, R; Lopez-Guajardo, C. C.; Brose, H. L.; Porter, K. I.; Leonard, R. A.; Hitt, A. A.; Schommer, S. L.; Elegbede, A. F.; Gould, M. N. (2003). "Production of knockout rats using ENU mutagenesis and a yeast-based screening assay". Nature Biotechnology. 21 (6): 645–51. doi:10.1038/nbt830. PMID 12754522.

209. Lee, R.C.; Ambros, V. (1993). "The C. elegans heterochronic gene lin-4 encodes small RNAs with antisense complementarity to lin-14". Cell. 75 (5): 843–854. doi:10.1016/0092-8674(93)90529-y. PMID 8252621.

210. Fire, A.; Xu, S.; Montgomery, M. K.; Kostas, S. A.; Driver, S. E.; Mello, C. C. (1998). "Potent and specific genetic interference by double-stranded RNA in Caenorhabditis elegans". Nature. 391 (6669): 806–811. doi:10.1038/35888. PMID 9486653.

211. Schwab, Rebecca; Ossowski, Stephan; Warthmann, Norman; Weigel, Detlef (2010-01-01). "Directed Gene Silencing with Artificial MicroRNAs". In Meyers, Blake C.; Green, Pamela J. Plant MicroRNAs. Methods in Molecular Biology. 592. Humana Press. pp. 71–88. doi:10.1007/978-1-60327-005-2_6. ISBN 9781603270045.

212. Vaucheret, H.; Chupeau, Y. (2011). "Ingested plant miRNAs regulate gene expression in animals". Cell Research. 22 (1): 3–5. doi:10.1038/cr.2011.164. PMC 3351922 . PMID 22025251.

213. Gentner, B.; Naldini, L. (2012-11-01). "Exploiting microRNA regulation for

genetic engineering". Tissue Antigens. 80 (5): 393–403. doi:10.1111/tan.12002. PMID 23020307.

214. Ladisch, M. R.; Kohlmann, K. L. (1992). "Recombinant human insulin". Biotechnology Progress. 8 (6): 469–478. doi:10.1021/bp00018a001. PMID 1369033.

215. Woodard, S. L.; Woodard, J. A.; Howard, M. E. (2004). "Plant molecular farming: Systems and products". Plant Cell Reports. 22 (10): 711–720. doi:10.1007/s00299-004-0767-1. PMID 14997337.

216. Ye, Xudong; Al-Babili, Salim; Klöti, Andreas; Zhang, Jing; Lucca, Paola; Beyer, Peter; Potrykus, Ingo (2000-01-14). "Engineering the Provitamin A (β-Carotene) Biosynthetic Pathway into (Carotenoid-Free) Rice Endosperm". Science. 287 (5451): 303–305. doi:10.1126/science.287.5451.303. PMID 10634784.

217. Peters, Pamela. "Transforming Plants – Basic Genetic Engineering Techniques". Retrieved 28 January 2010.

218. Voiland, Michael; McCandless, Linda (February 1999). "Development Of The "Gene Gun" At Cornell". Archived from the original on May 1, 2008.

219. Segelken, Roger (May 14, 1987). "Biologists Invent Gun for Shooting Cells with DNA Issue" (PDF). Cornell Chronicle. 18 (33): 3.

220. Timelines: 1987: Next The gene gun. lifesciencesfoundation.org

221. Clough, S. J.; Bent, A. F. (1998). "Floral dip: A simplified method for Agrobacterium-mediated transformation ofArabidopsis thaliana". The Plant Journal. 16 (6): 735–743. doi:10.1046/j.1365-313x.1998.00343.x. PMID 10069079.

222. Brophy, B.; Smolenski, G.; Wheeler, T.; Wells, D.; l'Huillier, P.; Laible, G. T. (2003). "Cloned transgenic cattle produce milk with higher levels of β-casein and κ-casein". Nature Biotechnology. 21 (2): 157–162. doi:10.1038/nbt783. PMID 12548290.

223. Hammer, R. E.; Pursel, V. G.; Rexroad, C. E.; Wall, R. J.; Bolt, D. J.; Ebert, K. M.; Palmiter, R. D.; Brinster, R. L. (1985). "Production of transgenic rabbits, sheep and pigs by microinjection". Nature. 315 (6021): 680–683. doi:10.1038/315680a0. PMID 3892305.

224. Clark, A. John (1998). "The Mammary Gland as a Bioreactor: Expression, Processing, and Production of Recombinant Proteins". Journal of Mammary Gland Biology and Neoplasia. 3 (3): 337–350. doi:10.1023/a:1018723712996. PMID 10819519.

225. Gordon, Katherine; Lee, Eric; Vitale, James A.; Smith, Alan E.; Westphal, Heiner; Hennighausen, Lothar (1987). "Production of human tissue plasmnogen activator in transgenic mouse milk". Biotechnology. 5 (11): 1183–1187. doi:10.1038/nbt1187-1183.

226. Gibson, D. G.; Glass, J. I.; Lartigue, C.; Noskov, V. N.; Chuang, R.-Y.; Algire, M. A.; Benders, G. A.; Montague, M. G.; Ma, L.; Moodie, M. M.; Merryman, C.; Vashee, S.; Krishnakumar, R.; Assad-Garcia, N.; Andrews-Pfannkoch, C.; Denisova, E. A.; Young, L.; Qi, Z.-Q.; Segall-Shapiro, T. H.; Calvey, C. H.;

Parmar, P. P.; Hutchison Ca, C. A.; Smith, H. O.; Venter, J. C. (2010). "Creation of a Bacterial Cell Controlled by a Chemically Synthesized Genome". Science. 329 (5987): 52–6. doi:10.1126/science.1190719. PMID 20488990.

227. Sample, Ian (20 May 2010). "Craig Venter creates synthetic life form". London: guardian.co.uk.
228. Malyshev, Denis A.; Dhami, Kirandeep; Lavergne, Thomas; Chen, Tingjian; Dai, Nan; Foster, Jeremy M.; Corrêa, Ivan R.; Romesberg, Floyd E. (2014-05-15). "A semi-synthetic organism with an expanded genetic alphabet". Nature. 509 (7500): 385–388. doi:10.1038/nature13314. PMC 4058825 . PMID 24805238.
229. Thyer, Ross; Ellefson, Jared (2014-05-15). "Synthetic biology: New letters for life's alphabet". Nature. 509 (7500): 291–292. doi:10.1038/nature13335.
230. Talbot, David (March 2016). "10 Breakthrough Technologies 2016: Precise Gene Editing in Plants". MIT Technology Review. Retrieved 2016-03-08.
231. Goeddel, D. V.; Kleid, D. G.; Bolivar, F.; Heyneker, H. L.; Yansura, D. G.; Crea, R.; Hirose, T.; Kraszewski, A.; Itakura, K.; Riggs, A. D. (1979). "Expression in Escherichia coli of chemically synthesized genes for human insulin". Proceedings of the National Academy of Sciences. 76 (1): 106–10. doi:10.1073/pnas.76.1.106. PMC 382885 . PMID 85300.
232. US Supreme Court Cases from Justia & Oyez (16 June 1980). "Diamond V Chakrabarty". 447 (303). Supreme.justia.com. Retrieved 17 July 2010.
233. "Artificial Genes". TIME. 15 November 1982. Retrieved 17 July 2010.
234. Bratspies, Rebecca (2007). "Some Thoughts on the American Approach to Regulating Genetically Modified Organisms" (PDF). Kansas Journal of Law and Public Policy. 16: 393.
235. BBC News (14 June 2002) GM crops: A bitter harvest?
236. Maugh, Thomas H. (June 9, 1987). Altered Bacterium Does Its Job : Frost Failed to Damage Sprayed Test Crop, Company Says. Los Angeles Times
237. Fraley, R. T.; Rogers, S. G.; Horsch, R. B.; Sanders, P. R.; Flick, J. S.; Adams, S. P.; Bittner, M. L.; Brand, L. A.; Fink, C. L.; Fry, J. S.; Galluppi, G. R.; Goldberg, S. B.; Hoffmann, N. L.; Woo, S. C. (1983). "Expression of bacterial genes in plant cells". Proceedings of the National Academy of Sciences of the United States of America. 80 (15): 4803–7. doi:10.1073/pnas.80.15.4803. PMC 384133 . PMID 6308651.
238. James, Clive (1996). "Global Review of the Field Testing and Commercialization of Transgenic Plants: 1986 to 1995" (PDF). The International Service for the Acquisition of Agri-biotech Applications. Retrieved 17 July 2010.
239. Vaeck, Mark; Reynaerts, Arlette; Höfte, Herman; Jansens, Stefan; De Beuckeleer, Marc; Dean, Caroline; Zabeau, Marc; Montagu, Marc Van; Leemans, Jan (1987). "Transgenic plants protected from insect attack". Nature. 328 (6125): 33–37. Bibcode:1987Natur.328...33V. doi:10.1038/328033a0.
240. "FDA Approves 1st Genetically Engineered Product for Food". Los Angeles Times. 24 March 1990. Retrieved 1 May 2014.

241. National Centre for Biotechnology Education (2006). Case Study: Chymosin. ncbe.reading.ac.uk
242. Campbell-Platt, Geoffrey (26 August 2011). Food Science and Technology. John Wiley & Sons. ISBN 978-1-4443-5782-0.
243. James, Clive (1997). “Global Status of Transgenic Crops in 1997”(PDF). ISAAA Briefs No. 5.: 31.
244. Bruening, G.; Lyons, J. M. (2000). “The case of the FLAVR SAVR tomato”. California Agriculture. 54 (4): 6–7. doi:10.3733/ca.v054n04p6.
245. MacKenzie, Debora (18 June 1994). “Transgenic tobacco is European first”. New Scientist.
246. Genetically Altered Potato Ok’d For Crops. Lawrence Journal (6 May 1995)
247. Global Status of Commercialized Biotech/GM Crops: 2011 ISAAA Brief ISAAA Brief 43-2011. Retrieved 14 October 2012
248. Pollack, Andrew (19 June 2013). “Executive at Monsanto wins global food honor”. The New York Times. Retrieved 20 June 2013.
249. Vàzquez-Salat, Núria; Salter, Brian; Smets, Greet; Houdebine, Louis-Marie (2012-11-01). “The current state of GMO governance: Are we ready for GM animals?”. Biotechnology Advances. Special issue on ACB 2011. 30 (6): 1336–1343. doi:10.1016/j.biotechadv.2012.02.006. PMID 22361646.
250. “AQUABOUNTY CLEARED TO SELL SALMON IN USA FOR COMMERCIAL PURPOSES”.
251. Bodnar, Anastasia (October 2010). “Risk Assessment and Mitigation of AquAdvantage Salmon” (PDF). ISB News Report.
252. Randerson, James (January 15, 2008) Arpad Pusztai: Biological divide. The Guardian
253. Waltz, Emily (2009-09-02). “GM crops: Battlefield”. Nature News. 461 (7260): 27–32. doi:10.1038/461027a. PMID 19727179.
254. “Rat study sparks GM furore”. Nature News & Comment.

CHAPTER 6

Genetically Modified Tree

Chapter Contents

- Introduction
- Optimisation
- Genetically Modified Tree
- Current Uses

INTRODUCTION

Tree breeding is the application of genetic, reproductive biology and economics principles to the genetic improvement and management of forest trees. In contrast to the selective breeding of livestock, arable crops, and horticultural flowers over the last few centuries, the breeding of trees, with the exception of fruit trees, is a relatively recent occurrence.

A typical forest tree breeding program starts with selection of superior phenotypes (plus trees) in a natural or planted forest. This application of mass selection improves the mean performance of the forest. Offspring is obtained from selected trees and grown in test plantations that act as genetic trials. Based on such tests the best genotypes among the parents can be selected. Selected trees are typically multiplied by either seeds or grafting and seed orchards are established when the preferred output is improved seed. Alternatively, the best genotypes can be directly propagated by cuttings or in-vitro methods and used directly in clonal plantations. The first system is frequently used in pines and other conifers, while the second is typical in some broadleaves (poplars, eucalypts and others). The objectives of a tree breeding program range from yield improvement and adaptation to particular conditions, to pest- and disease-resistance, wood properties, etc. Currently, tree breeding starting to take advantage of the fast development in plant genetics and genomics.

OPTIMISATION

Tree breeders make efforts to get their operation efficient by optimising tree breeding. Scientists develop tools aimed at improvement of the efficiency of tree breeding programmes. Optimising can mean adapting strategies and methods to certain species, group of populations, structure of genetic variation and mode of inheritance of the important traits to obtain the highest benefit per unit of time. Optimising is usually carried out at the following levels: breeding strategy (appropriate intensity of breeding, breeding population structure and size, plan for maintenance of genetic diversity), breeding methods (mating type, testing and selection methods, testing population size and time) and deployment methods of the genetically improved material (seed orchards and clonal forestry: genetic contribution, size). Computer simulators are frequently used: stochastic – based on defined and random algorithms; and deterministic – based on defined algorithms.

Selection strategies have been compared for annual progress in long-term breeding at a given annual cost considering genetic gain, gene diversity, cost components, and time components. For Norway spruce it seems favourable to clone full sib families and then select based on clonal performance while for Scots pine a two-stage strategy seems best, first phenotypic pre-selection and then progeny-testing the selections.

Tree Improvement

A genetically variable population and a method of selecting genetically superior individuals provide the basis for tree improvement by breeding. In essence, a tree improvement program sets out to isolate and evaluate the genetic component of variation in one or more characters of interest. In the simplest procedure, cycles of selection reduce the available population in a particular direction to enhance desirable traits, then breeding from selections to expand the population with improved characteristics. Breeding strategies vary with species and objectives, but all use mating designs to generate information and new material. Choice of a suitable breeding strategy and mating design is a key decision in any breeding program. Kiss (1986) used a 2-level design in British Columbia to study variation within and between separate populations of white spruce, both within British Columbia and from eastern North America.

The breeding program for white spruce initiated in 1986 by the Canadian Forestry Service in the Maritimes employed 2 kinds of mating: polycross, to test clones for general combining ability; and pair-mating, to generate material for second generation selections (Fowler et al. 1988).

Newton's (2003) systematic review of yield responses of white spruce and 3 other North American conifers to forest tree improvement practices indicated that correct provenance-progeny selection could yield juvenile height growth gains of about 12% at 20 years for white spruce, and a corresponding merchantable productivity (mean annual merchantable volume increment) gain of 26% at 50 years for plantations established at nominal initial densities on medium-to-good quality sites. Also, preliminary estimates derived from individual case studies indicated that first generational selection strategies for white spruce could increase merchantable productivity by approximately 20% at 45 years.

GENETICALLY MODIFIED TREE

A genetically modified tree (GMt, GM tree, genetically engineered tree, GE tree or transgenic tree) is a tree whose DNA has been modified using genetic engineering techniques. In most cases the aim is to introduce a novel trait to the plant which does not occur naturally within the species. Examples include resistance to certain pests, diseases, environmental conditions, and herbicide tolerance, or the alteration of lignin levels in order to reduce pulping costs.

Genetically modified forest trees are not yet approved ("deregulated") for commercial use, with the exception of insect-resistant poplar trees in China. and one case of GM Eucalyptus in Brazil. Several genetically modified forest tree species are undergoing field trials for deregulation, and much of the research is being carried out by the pulp and paper industry, primarily with the intention of increasing the productivity of existing tree stock. Certain genetically modified orchard tree species have been deregulated for commercial use in the United States including the papaya and plum. The development, testing and use of GM trees remains at an early stage in comparison to GM crops.

Technician checks on genetically modified peach and apple "orchards". Each dish holds experimental trees grown from lab-cultured cells to which researchers have given new genes. Source: USDA.

Research

Research into genetically modified trees has been ongoing since 1988. Concerns surrounding the biosafety implications of releasing genetically modified trees into the wild have held back regulatory approval of GM forest trees. This concern is exemplified in the Convention on Biological Diversity's stance:

The Conference of the Parties, Recognising the uncertainties related to the potential environmental and socio-economic impacts, including long term and trans-boundary impacts, of genetically modified trees on global forest biological diversity, as well as on the livelihoods of indigenous and local communities, and given the absence of reliable data and of capacity in some countries to undertake risk assessments and to evaluate those potential impacts, recommends parties to take a precautionary approach when addressing the issue of genetically modified trees.

A precondition for further commercialization of GM forest trees is likely to be their complete sterility. Plantation trees remain phenotypically similar to their wild cousins in that most are the product of no more than three generations of artificial selection, therefore, the risk of transgene escape by pollination with compatible wild species is high. One of the most credible science-based concerns with GM trees is their potential for wide dispersal of seed and pollen. The fact that pine pollen travels long distances is well established, moving up to 3,000 kilometers from its source. Additionally, many tree species reproduce for a long time before being harvested. In combination these factors have led some to believe that GM trees are worthy of special environmental considerations over GM crops. Ensuring sterility for GM trees has proven elusive, but efforts are being made. While tree geneticist Steve Strauss predicted that complete containment might be possible by 2020, many questions remain.

Proposed Uses

GM trees under experimental development have been modified with traits intended to provide benefit to industry, foresters or consumers. Due to high regulatory and research costs, the majority of genetically modified trees in silviculture consist of plantation trees, such as eucalyptus, poplar, and pine.

Lignin Alteration

Several companies in the pulp and paper industry are interested in utilizing GM technology to alter the lignin content of plantation trees. It is estimated that reducing lignin in plantation trees by genetic modification could reduce pulping costs by up to $15 per cubic metre. Lignin removal from wood fibres conventionally relies on costly and environmentally hazardous chemicals. By developing low-lignin GM trees it is hoped that pulping and bleaching processes will require fewer inputs, therefore, mills supplied by low-lignin GM trees may have a reduced impact on their surrounding ecosystems and communities. However, it is argued that reductions in lignin may compromise the structural integrity of the plant, thereby making it more susceptible to pathogens and disease, which could necessitate pesticide use exceeding that on traditional plantations.

Frost Tolerance

Genetic modification can allow trees to cope with abiotic stresses such that their geographic range is broadened. Freeze-tolerant GM eucalyptus trees for use in southern US plantations are currently being tested in open air sites with such an objective in mind. ArborGen, a tree biotechnology company and joint venture of pulp and paper firms Rubicon (New Zealand), MeadWestvaco (US) and International Paper (US) is leading this research. Until now the cultivation of eucalyptus has only been possible on the southern tip of Florida, freeze-tolerance would substantially extend the cultivation range northwards.

Accelerated Growth

In Brazil, field trials of fast growing GM eucalyptus are currently underway, they are set to conclude in 2015-2016 with commercialization to result. *FuturaGene*, a biotechnology company owned by Suzano, a Brazilian pulp and paper company, has been leading this research. Stanley Hirsch, chief executive of FuturaGene has stated: "Our trees grow faster and thicker. We are ahead of everyone. We have shown we can increase the yields and growth rates of trees more than anything grown by traditional breeding." The company is looking to reduce harvest cycles from 7 to 5.5 years with 20-30% more mass than conventional eucalyptus. There is concern that such objectives may further exacerbate the negative impacts of plantation forestry. Increased water and soil nutrient demand from faster growing species may lead to irrecoverable losses in site productivity and further impinge upon neighbouring communities and ecosystems.

Disease Resistance

Ecologically motivated research into genetic modification is underway. There are ongoing schemes that aim to foster disease resistance in trees such as the American chestnut (see Chestnut blight) and the English elm (see Dutch elm disease) for the purpose of their reintroduction to the wild. Specific diseases have reduced the populations of these emblematic species to the extent that they are mostly lost in the wild. Genetic modification is being pursued concurrently with traditional breeding techniques in an attempt to endow these species with disease resistance.

CURRENT USES

Poplars in China

In 2002 China's State Forestry Administration approved GM poplar trees for commercial use. Subsequently 1.4 million Bt (insecticide) producing GM poplars were planted in China. They were planted both for their wood and as part of China's 'Green Wall' project, which aims to impede desertification. Reports indicate that the GM poplars have spread beyond the area of original planting and that contamination of native poplars with the Bt gene is occurring. There is concern with these developments, particularly because the pesticide producing trait may impart a positive selective advantage on the poplar, allowing it a high level of invasiveness.

REFERENCES

1. Lindgren, D. and Prescher, F. 2005. Optimal clone number for seed orchards with tested clones. Silvae Genetica 54: 80-92.
2. Danusevičius D & Lindgren D 2002. Efficiency of Selection Based on Phenotype, Clone and Progeny Testing in Long-term Breeding. Silvae Genetica 51:19-26.
3. Danusevičius D & Lindgren D 2002. Two-stage selection strategies in tree breeding considering gain, diversity, time and cost. Forest Genetics. 9:145-157.
4. Kiss, G.K. 1986. Genetic improvement of white and Engelmann spruce in British Columbia 1983–85. p. 191–193 in Yeatman, C.W.; Boyle, T.J.B. (Eds.), Proc. 20th Meet. Can. Tree Improv. Assoc. Part 1, Quebec QC.
5. Fowler, D.P., Bonga, J.M., Park, Y.S., Simpson, J.D., and Smith, R.F. 1988. Tree breeding at the Canadian Forestry Service – Maritimes 1985 and 1986. p. 31–36 in Morgenstern, E.K.; Boyle, T.J.B. (Eds.). Tree Improvement – Progressing Together Sympos., Truro NS, Aug. 1987. Proc. Part 1, 21st Meet. Can. Tree Improv. Assoc.
6. Newton, P.F. 2003. Systematic review of yield responses of four North American conifers to forest tree improvement practices. For. Ecol. Manage. 172:29–51.
7. Wang, H. (2004). "The state of genetically modified forest trees in China" (PDF). Preliminary review of biotechnology in forestry, including genetic modification, Forest Genetic Resources Working Paper Forest Resources Development Service, Forest Resources Division. Rome, Italy. FAO: 96.
8. Sedjo, R.A. (2005). "Will Developing Countries be the Early Adopters of Genetically Engineered Forests?" (PDF). AgBioForum. 8 (4): 205.
9. "Brazil approves transgenic eucalyptus". Nature Biotechnology. 33 (6): 577–577. 9 June 2015. doi:10.1038/nbt0615-577c.
10. Sedjo, R.A. (2010). "Transgenic Trees for Biomass: The Effects of Regulatory Restrictions and Court Decisions on the Pace of Commercialization" (PDF). AgBioForum. 13 (4): 391.
11. Sedjo, R.A. (2010). "Transgenic Trees for Biomass: The Effects of Regulatory Restrictions and Court Decisions on the Pace of Commercialization" (PDF). AgBioForum. 13 (4): 393.
12. Kanowski, Peter. "Genetically-modified trees: opportunities for dialogue A scoping paper for The Forests Dialogue" (PDF). The Forest Dialogue. Retrieved 16 January 2014.
13. Walter, C. (2010). "The 20-year environmental safety record of GM trees" (PDF). Nature. 28 (7): 656–658. doi:10.1038/nbt0710-656.
14. "COP 8 Decision VIII/19 Forest biological diversity: implementation of the programme of work". Convention on Biological Diversity. Retrieved 16 January

2014.

15. Sedjo, R.A. (2004). "Genetically Engineered Trees: Promise and Concerns" (PDF). Resources for the Future: 20–21.
16. Bradshaw, A.H. (2001). "Plotting a course for GM forestry". Nature. 19 (12): 1103–1104. doi:10.1038/nbt1201-1103b.
17. Strauss, S.H. (2009). "Strangled at birth? Forest biotech and the Convention on Biological Diversity" (PDF). Nature Biotechnology. 27 (6): 524. doi:10.1038/nbt0609-519. PMID 19513052.
18. Williams, C.G. (2010). "Long-distance pine pollen still germinates after meso-scale dispersal" (PDF). American Journal of Botany. 97 (5): 846–855. doi:10.3732/ajb.0900255.
19. Kuparinen, A. (2008). "Assessing the risk of gene flow from genetically modified trees carrying mitigation transgenes" (PDF). Biological Invasions. 10: 282. doi:10.1007/s10530-007-9129-6.
20. James, R.R. (1997). "Utilizing a social ethic toward the environment in assessing genetically engineered insect-resistance in trees". Agriculture and Human Values. 14: 237–249. doi:10.1023/A:1007408811726.
21. Ahuja, M.R. (2011). "Fate of transgenes in the forest tree genome"(PDF). Tree Genetics & Genomes. 7: 226. doi:10.1007/s11295-010-0339-1.
22. "USDA Weighs Plan to Bring GM Eucalyptus to Southeast Pinelands". New York Times. January 29, 2010.
23. Barker, D.; et al. (2013). "Genetically Engineered Trees: The New Frontier of Biotechnology" (PDF). Center for Food Safety: 26.
24. Sedjo, R.A. (2004). "Genetically Engineered Trees: Promise and Concerns" (PDF). Resources for the Future: 15.
25. Owusu, R.A. (1999). "GM technology in the forest sector - A scoping study for WWF" (PDF). WWF: 10.
26. Nottingham, S. (2002). Genescapes - The Ecology of Genetic Engineering. Zed Books. ISBN 978-1842770375.
27. Doering, D. S. (2001). "Will the Marketplace See the Sustainable Forest for the Transgenic Trees?" (PDF). Proceedings of the First International Symposium on Ecological and Societal Aspects of Transgenic Plantations: 70–81. The communities at or near the plantations and the paper mills may receive a net environmental benefit of cleaner water and air in their communities. (p. 73)
28. Meilan, R. (2007). "Manipulating Lignin Biosynthesis to Improve Populus as a Bio-Energy Feedstock" (PDF). Institute of Forest Biotechnology, Genetically Engineered Forest Trees - Identifying Priorities for Ecological Risk Assessment: 55–61. Some scientists believe ... that reducing lignin content may lead to increases in cellulose content. But critics argue that reductions in lignin will compromise the structural integrity of the plant and make it more susceptible to pathogens and diseases. (p. 59)
29. Hall, C. (2007). "GM technology in forestry: lessons from the GM food

'debate'". International Journal of Biotechnology. 9 (5): 436–447. doi:10.1504/ijbt.2007.014270. Altering the quality or quantity of lignin may have significant impacts on the survival abilities of the tree, such as impairing its pest or disease resistance and necessitating the use of additional pesticides.

30. Mathews, J.H.; Campbell, M.M. (2000). "The advantages and disadvantages of the application of genetic engineering to forest trees: a discussion" (PDF). Forestry. 73 (4): 371–380. doi:10.1093/forestry/73.4.371. As Pullman et al.(1998) pointed out, modification of trees' adaptation to environmental stresses will enable foresters to grow more desirable commercial tree species on a broader range of soil types and planting sites. (p.375)
31. Barker, D.; et al. (2013). "Genetically Engineered Trees: The New Frontier of Biotechnology" (PDF). Center for Food Safety: 3.
32. Harfouche, A.; et al. (2011). "Tree genetic engineering and applications to sustainable forestry and biomass production". Trends in Biotechnology. 29 (1): 9–17. doi:10.1016/j.tibtech.2010.09.003. ArborGen is a joint venture between International Paper Company (USA) MeadWestvaco (USA) and Rubicon Limited (New Zealand) (p.13)
33. Institute of Forest Biotechnology (2007). "Genetically Engineered Forest Trees - Identifying Priorities for Ecological Risk Assessment - Summary of a Multistakeholder Workshop" (PDF). private company ArborGen is reportedly focusing on the development of three GE varieties: fast-growing loblolly pine for Southern pine plantations, low-lignin eucalyptus for use in South America, and cold-hardy eucalyptus for the Southern U.S. (p. ix)
34. "Deliberate release of genetically modified trees An abundance of poplars". GMO Safety. June 1, 2012. A gene has been introduced into the trees that makes them less sensitive to cold. Until now cultivation of eucalyptus in the US was only possible on the southern tip of Florida; frost tolerance could mean that cultivation would be possible in other parts of the USA.
35. Overbeek W. (2012). "An overview of industrial tree plantation conflicts in the global South. Conflicts, trends, and resistance struggles." (PDF). EJOLT. 3: 84.
36. Vidal, J. (November 15, 2012). "The GM tree plantations bred to satisfy the world's energy needs - Israeli biotech firm says its modified eucalyptus trees can displace the fossil fuel industry". Guardian.
37. Gerber, J.F. (2011). "Conflicts over industrial tree plantations in the South: Who, how and why?". Global Environmental Change. 21: 165–176. doi:10.1016/j.gloenvcha.2010.09.005. Fast-wood plantations tend to destabilize water cycles provoking reduced water flow throughout the year, the disappearance of streams during the dry season, and damages to other (agro-)ecosystems (p.167)
38. Owusu, R.A. (1999). "GM technology in the forest sector - A scoping study for WWF" (PDF). WWF. Biotechnology may inadvertently become yet another driver for inappropriate plantation development. Increased soil nutrient and water demand of fast growing species on short rotations could lead to irrecoverable loss of site productivity. (p.5)

39. Nottingham, S. (2002). Genescapes - The Ecology of Genetic Engineering. Zed Books. fast-growing transgenic trees will make additional demands on soil nutrients and water, with consequences for the long-term fertility of soils. Substantial fertilizer inputs might be necessary to maintain high yields
40. "Into the Wildwood". The Economist. May 4, 2013.
41. Harfouche, A. (2011). "Tree genetic engineering and applications to sustainable forestry and biomass production" (PDF). Trends in Biotechnology. 29 (1): 13. doi:10.1016/j.tibtech.2010.09.003.
42. Powell, William (march 2014) "the American Chestnut's Genetic Rebirth", Scientific American, Volume 310, Number 3, Page 52
43. Lang, Chris (2004). "China: Genetically modified madness". The World Rainforest Movement. Retrieved 29 January 2014. Two years ago, China's State Forestry Administration approved genetically modified (GM) poplar trees for commercial planting.
44. Then, C.; Hamberger, S. (2010). "Genetically engineered trees – a ticking "time bomb"?" (PDF). Testbiotech.
45. Sedjo, R.A. (2005). "Will Developing Countries be the Early Adopters of Genetically Engineered Forests? Resources for the Future" (PDF). AgBioForum. 8 (4): 205–211. the engineered gene has probably spread beyond the area of the original plantings (p.206)
46. Carman, N. (2006). "Ecological and Social Impacts of Fast Growing Timber Plantations and Genetically Engineered Trees" (PDF). Dogwood Alliance. The Nanjing Institute of Environmental Science has reported that contamination of native poplars with the Bt gene is already occurring. (p.4)
47. Then, C.; Hamberger, S. (2010). "Genetically engineered trees – a ticking "time bomb"?" (PDF). Testbiotech. Bt poplars are grown alongside non-transgenic trees, possibly delaying the emergence of resistances. If this is the case, the transgenic poplars will have a higher fitness in comparison to the other trees, thus conceivably fostering their invasiveness in the mid or even long term. (p.16)

CHAPTER 7

Genetically Modified Crops

Chapter Contents

- **Introduction**
- **Gene Transfer in Nature and Traditional Agriculture**
- **Methods**
- **Types of Modifications**
- **Economics**
- **Farming Practices**

The repeated definition of GM trees, crops, fruits etc tend to create an impression of pseudo monotony which a careful rephrasing and rearrangement could minimize because what passes for some trees may generally go for others in the family of crops. Collectively, these techniques are known as recombinant DNA technology. Other terms used for GM plants or foods derived from them are genetically modified organism (GMO), genetically engineered (GE), bioengineered, and transgenic. 'Genetically modified' is an imprecise term and a potentially confusing one, in that virtually everything we eat has been modified genetically through domestication from wild species and many generations of selection by humans for desirable traits. The term is used here because it is the one most widely used to indicate the use of recombinant DNA technology. According to USDA standards for organic agriculture, seeds or other substances derived through GM technology are not allowed in organic production.

INTRODUCTION

Genetically modified crops (GMCs, GM crops, or biotech crops) are plants used in agriculture, the DNA of which has been modified using genetic engineering techniques. In most cases, the aim is to introduce a new trait to the plant which does not occur naturally in the species. Examples in food crops include resistance to certain pests, diseases, or environmental conditions, reduction of spoilage, or resistance to chemical treatments (e.g. resistance to a herbicide), or improving the nutrient profile of the crop. Examples in non-food crops include production of pharmaceutical agents, biofuels, and other industrially useful goods, as well as for bioremediation.

Farmers have widely adopted GM technology. Between 1996 and 2015, the total surface area of land cultivated with GM crops increased by a factor of 100, from 17,000 km^2 (4.2 million acres) to 1,797,000 km^2 (444 million acres). 10% of the world's arable land was planted with GM crops in 2010. In the US, by 2014, 94% of the planted area of soybeans, 96% of cotton and 93% of corn were genetically modified varieties. Use of GM crops expanded rapidly in developing countries, with about 18 million farmers growing 54% of worldwide GM crops by 2013. A 2014 meta-analysis concluded that GM technology adoption had reduced chemical pesticide use by 37%, increased crop yields by 22%, and increased farmer profits by 68%. This reduction in pesticide use has been ecologically beneficial, but benefits may be reduced by overuse. Yield gains and pesticide reductions are larger for insect-resistant crops than for herbicide-tolerant crops. Yield and profit gains are higher in developing countries than in developed countries.

There is a scientific consensus that currently available food derived from GM crops poses no greater risk to human health than conventional food, but that each GM food needs to be tested on a case-by-case basis before introduction. Nonetheless, members of the public are much less likely than scientists to perceive GM foods as safe. The legal and regulatory status of GM foods varies by country, with some nations banning or restricting them, and others permitting them with widely differing degrees of regulation.

However, opponents have objected to GM crops on several grounds, including environmental concerns, whether food produced from GM crops is safe, whether GM crops are needed to address the world's food needs, and concerns raised by the fact these organisms are subject to intellectual property law.

GENE TRANSFER IN NATURE AND TRADITIONAL AGRICULTURE

DNA transfers naturally between organisms. Several natural mechanisms allow gene flow across species. These occur in nature on a large scale – for example, it is one mechanism for the development of antibiotic resistance in bacteria. This is facilitated by transposons, retrotransposons, proviruses and other mobile genetic elements that naturally translocate DNA to new loci in a genome. Movement occurs over an evolutionary time scale.

The introduction of foreign germplasm into crops has been achieved by traditional crop breeders by overcoming species barriers. A hybrid cereal grain was created in 1875, by crossing wheat and rye. Since then important traits including dwarfing genes and rust resistance have been introduced. Plant tissue culture and deliberate mutations have enabled humans to alter the makeup of plant genomes.

The first genetically modified crop plant was produced in 1982, an antibiotic-resistant tobacco plant. The first field trials occurred in France and the USA in 1986, when tobacco plants were engineered for herbicide resistance. In 1987, Plant Genetic Systems (Ghent, Belgium), founded by Marc Van Montagu and Jeff Schell, was the first company to genetically engineer insect-resistant (tobacco) plants by incorporating genes that produced insecticidal proteins from *Bacillus thuringiensis* (Bt).

The People's Republic of China was the first country to allow commercialized transgenic plants, introducing a virus-resistant tobacco in 1992,which was withdrawn in 1997.[3] The first genetically modified crop approved for sale in the U.S., in 1994, was the *FlavrSavr* tomato. It had a longer shelf life, because it took longer to soften after ripening. In 1994, the European Union approved tobacco engineered to be resistant to the herbicide bromoxynil, making it the first commercially genetically engineered crop marketed in Europe.

In 1995, Bt Potato was approved by the US Environmental Protection Agency, making it the country's first pesticide producing crop. In 1995 canola with modified oil composition (Calgene), Bt maize (Ciba-Geigy), bromoxynil-resistant cotton (Calgene), Bt cotton (Monsanto), glyphosate-resistant soybeans (Monsanto), virus-resistant squash (Asgrow), and additional delayed ripening tomatoes (DNAP, Zeneca/Peto, and Monsanto) were approved. As of mid-1996, a total of 35 approvals had been granted to commercially grow 8 transgenic crops and one flower crop (carnation), with 8 different traits in 6 countries plus the EU. In 2000, Vitamin A-enriched golden rice was developed, though as of 2016 it was not yet in commercial production. In 2013 the leaders of the three research teams that first applied genetic engineering to crops, Robert Fraley, Marc Van Montagu and Mary-Dell Chilton were awarded the World Food Prize for improving the "quality, quantity or availability" of food in the world.

METHODS

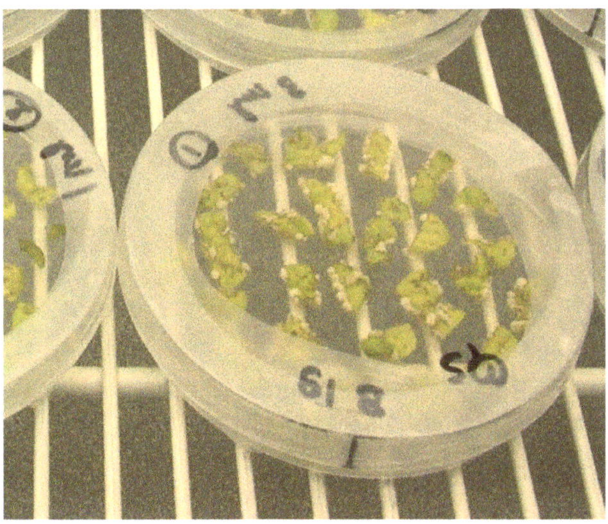

Plants (*Solanum chacoense*) being transformed using agrobacterium

Genetically engineered crops have genes added or removed using genetic engineering techniques,originally including gene guns, electroporation, microinjection and agrobacterium. More recently, CRISPR and TALEN offered much more precise and convenient editing techniques.

Gene guns (also known as biolistics) "shoot" (direct high energy particles or radiations against) target genes into plant cells. It is the most common method. DNA is bound to tiny particles of gold or tungsten which are subsequently shot into plant tissue or single plant cells under high pressure. The accelerated particles penetrate both the cell wall and membranes. The DNA separates from the metal and is integrated into plant DNA inside the nucleus. This method has been applied successfully for many cultivated crops, especially monocots like wheat or maize, for which transformation using *Agrobacterium tumefaciens* has been less successful. The major disadvantage of this procedure is that serious damage can be done to the cellular tissue.

Agrobacterium tumefaciens-mediated transformation is another common technique. Agrobacteria are natural plant parasites, and their natural ability to transfer genes provides another engineering method. To create a suitable environment for themselves, these Agrobacteria insert their genes into plant hosts, resulting in a proliferation of modified plant cells near the soil level (crown gall). The genetic information for tumour growth is encoded on a mobile, circular DNA fragment (plasmid). When *Agrobacterium* infects a plant, it transfers this T-DNA to a random site in the plant genome. When used in genetic engineering the bacterial T-DNA is removed from the bacterial plasmid and replaced with the desired foreign gene. The bacterium is a vector, enabling transportation of foreign genes into plants. This method works especially well for dicotyledonous plants like potatoes, tomatoes, and tobacco. Agrobacteria infection is less successful in crops like wheat and maize.

Electroporation is used when the plant tissue does not contain cell walls. In this technique, "DNA enters the plant cells through miniature pores which are temporarily caused by electric pulses."

Microinjection directly injects the gene into the DNA.

Plant scientists, backed by results of modern comprehensive profiling of crop composition, point out that crops modified using GM techniques are less likely to have unintended changes than are conventionally bred crops.

In research tobacco and *Arabidopsis thaliana* are the most frequently modified plants, due to well-developed transformation methods, easy propagation and well studied genomes. They serve as model organisms for other plant species.

Introducing new genes into plants requires a promoter specific to the area where the gene is to be expressed. For instance, to express a gene only in rice grains and not in leaves, an endosperm-specific promoter is used. The codons of the gene must be optimized for the organism due to codon usage bias.

TYPES OF MODIFICATIONS

Transgenic maize containing a gene from the bacteria *Bacillus thuringiensis*

Transgenic

Transgenic plants have genes inserted into them that are derived from another species. The inserted genes can come from species within the same kingdom (plant to plant) or between kingdoms (for example, bacteria to plant). In many cases the inserted DNA has to be modified slightly in order to correctly and efficiently express in the host organism. Transgenic plants are used to express proteins like the cry toxins from *B. thuringiensis*, herbicide resistant genes, antibodies and antigens for vaccinations A study led by the European Food Safety Authority (EFSA) found also viral genes in transgenic plants.

Transgenic carrots have been used to produce the drug Taliglucerase alfa which is used to treat Gaucher's disease. In the laboratory, transgenic plants have been modified to increase photosynthesis (currently about 2% at most plants versus the theoretic potential of 9–10%). This is possible by changing the rubisco enzyme (i.e. changing C3 plants into C4 plants), by placing the rubisco in a carboxysome, by adding CO_2 pumps in the cell wall, by changing the leaf form/size. Plants have been engineered to exhibit bioluminescence that may become a sustainable alternative to electric lighting.

Cisgenic

Cisgenic plants are made using genes found within the same species or a closely related one, where conventional plant breeding can occur. Some breeders and scientists argue that cisgenic modification is useful for plants that are difficult to crossbreed by conventional means (such as potatoes), and that plants in the cisgenic category should not require the same regulatory scrutiny as transgenics.

Subgenic

Genetically modified plants can also be developed using gene knockdown or gene knockout to alter the genetic makeup of a plant without incorporating genes from other plants. In 2014, Chinese researcher Gao Caixia filed patents on the creation of a strain of wheat that is resistant to powdery mildew. The strain lacks genes that encode proteins that repress defenses against the mildew. The researchers deleted all three copies of the genes from wheat's hexaploid genome. Gao used the TALENs and CRISPR gene editing tools without adding or changing any other genes. No field trials were immediately planned. The CRISPR technique has also been used to modify white button mushrooms (*Agaricus bisporus*).

ECONOMICS

GM food's economic value to farmers is one of its major benefits, including in developing nations. A 2010 study found that Bt corn provided economic benefits of $6.9 billion over the previous 14 years in five Midwestern states. The majority ($4.3 billion) accrued to farmers producing non-Bt corn. This was attributed to European corn borer populations reduced by exposure to Bt corn, leaving fewer to attack conventional corn nearby. Agriculture economists calculated that "world surplus increased by $240.3 million for 1996. Of this total, the largest share (59%) went to U.S. farmers. Seed company Monsanto received the next largest share (21%), followed by US consumers (9%), the rest of the world (6%), and the germplasm supplier, Delta & Pine Land Company of Mississippi (5%)."

According to the International Service for the Acquisition of Agri-biotech Applications (ISAAA), in 2014 approximately 18 million farmers grew biotech crops in 28 countries; about 94% of the farmers were resource-poor in developing countries. 53% of the global biotech crop area of 181.5 million hectares was grown in 20 developing countries. PG Economics comprehensive 2012 study concluded that GM crops increased farm incomes worldwide by $14 billion in 2010, with over half this total going to farmers in developing countries.

Critics challenged the claimed benefits to farmers over the prevalence of biased observers and by the absence of randomized controlled trials. The main Bt crop grown by small farmers in developing countries is cotton. A 2006 review of Bt cotton findings by agricultural economists concluded, "the overall balance sheet, though promising, is mixed. Economic returns are highly variable over years, farm type, and geographical location".

In 2013 the European Academies Science Advisory Council (EASAC) asked the EU to allow the development of agricultural GM technologies to enable more sustainable agriculture, by employing fewer land, water and nutrient resources. EASAC also criticizes the EU's "timeconsuming and expensive regulatory framework" and said that the EU had fallen behind in the adoption of GM technologies.

Participants in agriculture business markets include seed companies, agrochemical companies, distributors, farmers, grain elevators and universities that develop new crops/traits and whose agricultural extensions advise farmers on best practices. According to a 2012 review based on data from the late 1990s and early 2000s, much of the GM crop grown each year is used for livestock feed and increased demand for meat leads to increased demand for GM feedcrops. Feed grain usage as a percentage of total crop production is 70% for corn and more than 90% of oil seed meals such as soybeans. About 65 million metric tons of GM corn grains and about 70 million metric tons of soybean meals derived from GM soybean become feed.

In 2014 the global value of biotech seed was US$15.7 billion; US$11.3 billion (72%) was in industrial countries and US$4.4 billion (28%) was in the developing countries. In 2009, Monsanto had $7.3 billion in sales of seeds and from licensing its technology; DuPont, through its Pioneer subsidiary, was the next biggest company in that market. As of 2009, the overall Roundup line of products including the GM seeds represented about 50% of Monsanto's business.

Some patents on GM traits have expired, allowing the legal development of generic strains that include these traits. For example, generic glyphosate-tolerant GM soybean is now available. Another impact is that traits developed by one vendor can be added to another vendor's proprietary strains, potentially increasing product choice and competition. The patent on the first type of *Roundup Ready* crop that Monsanto produced (soybeans) expired in 2014 and the first harvest of off-patent soybeans occurs in the spring of 2015. Monsanto has broadly licensed the patent to other seed companies that include the glyphosate resistance trait in their seed products. About 150 companies have licensed the technology, including Syngenta and DuPont Pioneer.

Yield

In 2014, the largest review yet concluded that GM crops' effects on farming were positive. The meta-analysis considered all published English-language examinations of the agronomic and economic impacts between 1995 and March 2014 for three major GM crops: soybean, maize, and cotton. The study found that herbicide-tolerant crops have lower production costs, while for insect-resistant crops the reduced pesticide use was offset by higher seed prices, leaving overall production costs about the same.

Yields increased 9% for herbicide tolerance and 25% for insect resistant varieties. Farmers who adopted GM crops made 69% higher profits than those who did not. The review found that GM crops help farmers in developing countries, increasing yields by 14 percentage points.

The researchers considered some studies that were not peer-reviewed, and a few that did not report sample sizes. They attempted to correct for publication bias, by considering sources beyond academic journals. The large data set allowed the study to control for potentially confounding variables such as fertiliser use. Separately, they concluded that the funding source did not influence study results.

Traits

GM crops grown today, or under development, have been modified with various traits. These traits include improved shelf life, disease resistance, stress resistance, herbicide resistance, pest resistance, production of useful goods such as biofuel or drugs, and ability to absorb toxins and for use in bioremediation of pollution.

Recently, research and development has been targeted to enhancement of crops that are locally important in developing countries, such as insect-resistant cowpea for Africa and insect-resistant brinjal (eggplant).

Lifetime

The first genetically modified crop approved for sale in the U.S. was the *FlavrSavr* tomato, which had a longer shelf life. It is no longer on the market.

In November 2014, the USDA approved a GM potato that prevents bruising.

In February 2015 Arctic Apples were approved by the USDA, becoming the first genetically modified apple approved for US sale. Gene silencing was used to reduce the expression of polyphenol oxidase (PPO), thus preventing enzymatic browning of the fruit after it has been sliced open. The trait was added to Granny Smith and Golden Delicious varieties. The trait includes a bacterial antibiotic resistance gene that provides resistance to the antibiotic kanamycin. The genetic engineering involved cultivation in the presence of kanamycin, which allowed only resistant cultivars to survive. Humans consuming apples do not acquire kanamycin resistance, per arcticapple.com. The FDA approved the apples in March 2015.

Nutrition

Edible Oils

Some GM soybeans offer improved oil profiles for processing or healthier eating. Camelina sativa has been modified to produce plants that accumulate high levels of oils similar to fish oils.

Vitamin Enrichment

Golden rice, developed by the International Rice Research Institute (IRRI), provides greater amounts of Vitamin A targeted at reducing Vitamin A deficiency. As of January 2016, golden rice has not yet been grown commercially in any country.

Researchers vitamin-enriched corn derived from South African white corn variety M37W, producing a 169-fold increase in Vitamin A, 6-fold increase in Vitamin C and doubled concentrations of folate. Modified Cavendish bananas express 10-fold the amount of Vitamin A as unmodified varieties.

Toxin Reduction

A genetically modified cassava under development offers lower cyanogen glucosides and enhanced protein and other nutrients (called BioCassava).

In November 2014, the USDA approved a potato, developed by J.R. Simplot Company, that prevents bruising and produces less acrylamide when fried. The modifications prevent natural, harmful proteins from being made via RNA interference. They do not employ genes from non-potato species. The trait was added to the Russet Burbank, Ranger Russet and Atlantic varieties.

Stress Resistance

Plants engineered to tolerate non-biological stressors such as drought, frost, high soil salinity, and nitrogen starvation were in development. In 2011, Monsanto's DroughtGard maize became the first drought-resistant GM crop to receive US marketing approval.

Herbicides

Glyphosate

As of 1999 the most prevalent GM trait was glyphosate-resistance. Glyphosate, (the active ingredient in Roundup and other herbicide products) kills plants by interfering with the shikimate pathway in plants, which is essential for the synthesis of the aromatic amino acids phenylalanine, tyrosine and tryptophan. The shikimate pathway is not present in animals, which instead obtain aromatic amino acids from their diet. More specifically, glyphosate inhibits the enzyme 5-enolpyruvylshikimate-3-phosphate synthase (EPSPS).

This trait was developed because the herbicides used on grain and grass crops at the time were highly toxic and not effective against narrow-leaved weeds. Thus, developing crops that could withstand spraying with glyphosate would both reduce environmental and health risks, and give an agricultural edge to the farmer.

Some micro-organisms have a version of EPSPS that is resistant to glyphosate inhibition. One of these was isolated from an *Agrobacterium* strain CP4 (CP4 EPSPS) that was resistant to glyphosate. The CP4 EPSPS gene was engineered for plant expression by fusing the 5' end of the gene to a chloroplast transit peptide derived from the petunia EPSPS. This transit peptide was used because it had shown previously an ability to deliver bacterial EPSPS to the chloroplasts of other plants. This CP4 EPSPS gene was cloned and transfected into soybeans.

The plasmid used to move the gene into soybeans was PV-GMGTO4. It contained three bacterial genes, two CP4 EPSPS genes, and a gene encoding beta-glucuronidase (GUS) from *Escherichia coli* as a marker. The DNA was injected into the soybeans using the particle acceleration method. Soybean cultivar A5403 was used for the transformation.

Bromoxynil

Tobacco plants have been engineered to be resistant to the herbicide bromoxynil.

Glufosinate

Crops have been commercialized that are resistant to the herbicide glufosinate, as well. Crops engineered for resistance to multiple herbicides to allow farmers to use a mixed group of two, three, or four different chemicals are under development to combat growing herbicide resistance.

2,4-D

In October 2014 the US EPA registered Dow's Enlist Duo maize, which is genetically modified to be resistant to both glyphosate and 2,4-D, in six states. Inserting a bacterial aryloxyalkanoate dioxygenase gene, *aad1* makes the corn resistant to 2,4-D. The USDA had approved maize and soybeans with the mutation in September 2014.

Dicamba

Monsanto has requested approval for a stacked strain that is tolerant of both glyphosate and dicamba.

Pest Resistance

Insects

Tobacco, corn, rice and many other crops have been engineered to express genes encoding for insecticidal proteins from Bacillus thuringiensis (Bt). Papaya, potatoes, and squash have been engineered to resist viral pathogens such as cucumber mosaic virus which, despite its name, infects a wide variety of plants. The introduction of Bt crops during the period between 1996 and 2005 has been estimated to have reduced the total volume of insecticide active ingredient use in the United States by over 100 thousand tons. This represents a 19.4% reduction in insecticide use.

In the late 1990s, a genetically modified potato that was resistant to the Colorado potato beetle was withdrawn because major buyers rejected it, fearing consumer opposition.

Viruses

Virus resistant papaya were developed In response to a papaya ringspot virus (PRV) outbreak in Hawaii in the late 1990s. . They incorporate PRV DNA. By 2010, 80% of Hawaiian papaya plants were genetically modified.

Potatoes were engineered for resistance to potato leaf roll virus and Potato virus Y in 1998. Poor sales led to their market withdrawal after three years.

Yellow squash that were resistant to at first two, then three viruses were developed, beginning in the 1990s. The viruses are watermelon, cucumber and zucchini/courgette yellow mosaic. Squash was the second GM crop to be approved by US regulators. The trait was later added to zucchini.

Many strains of corn have been developed in recent years to combat the spread of Maize dwarf mosaic virus, a costly virus that causes stunted growth which is carried in Johnson grass and spread by aphid insect vectors. These strands are commercially available although the resistance is not standard among GM corn variants.

By-Products

Drugs

In 2012, the FDA approved the first plant-produced pharmaceutical, a treatment for Gaucher's Disease. Tobacco plants have been modified to produce therapeutic antibodies.

Biofuel

Algae is under development for use in biofuels. Researchers in Singapore were working on GM jatropha for biofuel production. Syngenta has USDA approval to market a maize trademarked Enogen that has been genetically modified to convert its starch to sugar for ethanol. In 2013, the Flemish Institute for Biotechnology was investigating poplar trees genetically engineered to contain less lignin to ease conversion into ethanol. Lignin is the critical limiting factor when using wood to make bio-ethanol because lignin limits the accessibility of cellulose microfibrils to depolymerization by enzymes.

Materials

Companies and labs are working on plants that can be used to make bioplastics. Potatoes that produce industrially useful starches have been developed as well. Oilseed can be modified to produce fatty acids for detergents, substitute fuels and petrochemicals.

Bioremediation

Scientists at the University of York developed a weed (*Arabidopsis thaliana*) that contains genes from bacteria that could clean TNT and RDX-explosive soil contaminants in 2011. 16 million hectares in the USA (1.5% of the total surface) are estimated to be contaminated with TNT and RDX. However *A. thaliana* was not tough enough for use on military test grounds. Modifications in 2016 included switchgrass and bentgrass.

Genetically modified plants have been used for bioremediation of contaminated soils. Mercury, selenium and organic pollutants such as polychlorinated biphenyls (PCBs).

Marine environments are especially vulnerable since pollution such as oil spills are not containable. In addition to anthropogenic pollution, millions of tons of petroleum annually enter the marine environment from natural seepages. Despite its toxicity, a considerable fraction of petroleum oil entering marine systems is eliminated by the hydrocarbon-degrading activities of microbial communities. Particularly successful is a recently discovered group of specialists, the so-called hydrocarbonoclastic bacteria (HCCB) that may offer useful genes.

Asexual Reproduction

Crops such as maize reproduce sexually each year. This randomizes which genes get propagated to the next generation, meaning that desirable traits can be lost. To maintain a high-quality crop, some farmers purchase seeds every year. Typically, the seed company maintains two inbred varieties, and crosses them into a hybrid strain that is then sold. Related plants like sorghum and gamma grass are able to perform apomixis, a form of asexual reproduction that keeps the plant's DNA intact. This trait is apparently controlled by a single dominant gene, but traditional breeding has been unsuccessful in creating asexually-reproducing maize. Genetic engineering offers another route to this goal. Successful modification would allow farmers to replant harvested seeds that retain desirable traits, rather than relying on purchased seed.

Crops

Herbicide Tolerance

GMO	Use	Countries approved in	First approved	Notes
Alfalfa	Animal feed	USA	2005	Approval withdrawn in 2007 and then re-approved in 2011
Canola	Cooking oil Margarine Emulsifiers in packaged foods	Australia	2003	
		Canada	1995	
		USA	1995	
Cotton	Fiber Cottonseed oil Animal feed	Argentina	2001	
		Australia	2002	
		Brazil	2008	
		Columbia	2004	
		Costa Rica	2008	
		Mexico	2000	
		Paraguay	2013	
		South Africa	2000	
		USA	1994	

GMO	Use	Countries approved in	First approved	Notes
Maize	Animal feed high-fructose corn syrup corn starch	Argentina	1998	
		Brazil	2007	
		Canada	1996	
		Colombia	2007	
		Cuba	2011	
		European Union	1998	Grown in Portugal, Spain, Czech Republic, Slovakia and Romania
		Honduras	2001	
		Paraguay	2012	
		Philippines	2002	
		South Africa	2002	
		USA	1995	
		Uruguay	2003	
Soybean	Animal feed Soybean oil	Argentina	1996	
		Bolivia	2005	
		Brazil	1998	
		Canada	1995	
		Chile	2007	
		Costa Rica	2001	
		Mexico	1996	
		Paraguay	2004	
		South Africa	2001	
		USA	1993	
		Uruguay	1996	
Sugar Beet	Food	Canada	2001	
		USA	1998	Commercialised 2007, production blocked 2010, resumed 2011.

Insect Resistance

GMO	Use	Countries approved in	First approved	Notes
Cotton	Fiber Cottonseed oil Animal feed	Argentina	1998	
		Australia	2003	
		Brazil	2005	
		Burkina Faso	2009	
		China	1997	
		Colombia	2003	
		Costa Rica	2008	
		India	2002	Largest producer of Bt cotton
		Mexico	1996	
		Myanmar	2006[N 1]	
		Pakistan	2010[N 1]	
		Paraguay	2007	
		South Africa	1997	
		Sudan	2012	
		USA	1995	

GMO	Use	Countries approved in	First approved	Notes
Eggplant	Food	Bangladesh	2013	12 ha planted on 120 farms in 2014
Maize	Animal feed high-fructose corn syrup corn starch	Argentina	1998	
		Brazil	2005	
		Columbia	2003	
		Mexico	1996	Centre of origin for maize
		Paraguay	2007	
		Philippines	2002	
		South Africa	1997	
		Uruguay	2003	
		USA	1995	
Poplar	Tree	China	1998	543 ha of bt poplar planted in 2014

Other Modified Traits

GMO	Use	Trait	Countries approved in	First approved	Notes
Canola	Cooking oil Margarine Emulsifiers in packaged foods	High laurate canola	Canada	1996	
			USA	1994	
		Phytase production	USA	1998	
Carnation	Ornamental	Delayed senescence	Australia	1995	
			Norway	1998	
		Modified flower colour	Australia	1995	
			Columbia	2000	In 2014 4 ha were grown in greenhouses for export
			European Union	1998	Two events expired 2008, another approved 2007
			Japan	2004	
			Malaysia	2012	For ornamental purposes
			Norway	1997	
Maize	Animal feed high-fructose corn syrup corn starch	Increased lysine	Canada	2006	
			USA	2006	
		Drought tolerance	Canada	2010	
			USA	2011	
Papaya	Food	Virus resistance	China	2006	
			USA	1996	Mostly grown in Hawaii
Petunia	Ornamental	Modified flower colour	China	1997	

GMO	Use	Trait	Countries approved in	First approved	Notes
Potato	Food	Virus resistance	Canada	1999	
			USA	1997	
	Industrial	Modified starch	USA	2014	
Rose	Ornamental	Modified flower colour	Australia	2009	Surrendered renewal
			Colombia	2010[N 2]	Greenhouse cultivation for export only.
			Japan	2008	
			USA	2011	
Soybean	Animal feed	Increased oleic acid production	Argentina	2015	
			Canada	2000	
			USA	1997	
	Soybean oil	Stearidonic acid production	Canada	2011	
			USA	2011	
Squash	Food	Virus resistance	USA	1994	
Sugar Cane	Food	Drought tolerance	Indonesia	2013	Environmental certificate only
Tobacco	Cigarettes	Nicotine reduction	USA	2002	

Development

The number of USDA-approved field releases for testing grew from 4 in 1985 to 1,194 in 2002 and averaged around 800 per year thereafter. The number of sites per release and the number of gene constructs (ways that the gene of interest is packaged together with other elements)—have rapidly increased since 2005. Releases with agronomic properties (such as drought resistance) jumped from 1,043 in 2005 to 5,190 in 2013. As of September 2013, about 7,800 releases had been approved for corn, more than 2,200 for soybeans, more than 1,100 for cotton, and about 900 for potatoes. Releases were approved for herbicide tolerance (6,772 releases), insect resistance (4,809), product quality such as flavor or nutrition (4,896), agronomic properties like drought resistance (5,190), and virus/fungal resistance (2,616). The institutions with the most authorized field releases include Monsanto with 6,782, Pioneer/DuPont with 1,405, Syngenta with 565, and USDA's Agricultural Research Service with 370. As of September 2013 USDA had received proposals for releasing GM rice, squash, plum, rose, tobacco, flax and chicory.

FARMING PRACTICES

Resistance

Bt Resistance

Constant exposure to a toxin creates evolutionary pressure for pests resistant to that toxin. Overreliance on glyphosate and a reduction in the diversity of weed management practices allowed the spread of glyphosate resistance in 14 weed species/biotypes in the US.

One method of reducing resistance is the creation of refuges to allow nonresistant organisms to survive and maintain a susceptible population.

To reduce resistance to Bt crops, the 1996 commercialization of transgenic cotton and maize came with a management strategy to prevent insects from becoming resistant. Insect resistance management plans are mandatory for Bt crops. The aim is to encourage a large population of pests so that any (recessive) resistance genes are diluted within the population. Resistance lowers evolutionary fitness in the absence of the stressor (Bt). In refuges, non-resistant strains outcompete resistant ones.

With sufficiently high levels of transgene expression, nearly all of the heterozygotes (S/s), i.e., the largest segment of the pest population carrying a resistance allele, will be killed before maturation, thus preventing transmission of the resistance gene to their progeny. Refuges (i. e., fields of nontransgenic plants) adjacent to transgenic fields increases the likelihood that homozygous resistant (s/s) individuals and any surviving heterozygotes will mate with susceptible (S/S) individuals from the refuge, instead of with other individuals carrying the resistance allele. As a result, the resistance gene frequency in the population remains lower.

Complicating factors can affect the success of the high-dose/refuge strategy. For example, if the temperature is not ideal, thermal stress can lower Bt toxin production and leave the plant more susceptible. More importantly, reduced late-season expression has been documented, possibly resulting from DNA methylation of the promoter. The success of the high-dose/refuge strategy has successfully maintained the value of Bt crops. This success has depended on factors independent of management strategy, including low initial resistance allele frequencies, fitness costs associated with resistance, and the abundance of non-Bt host plants outside the refuges.

Companies that produce Bt seed are introducing strains with multiple Bt proteins. Monsanto did this with Bt cotton in India, where the product was rapidly adopted. Monsanto has also; in an attempt to simplify the process of implementing refuges in fields to comply with Insect Resistance Management(IRM) policies and prevent irresponsible planting practices; begun marketing seed bags with a set proportion of refuge (non-transgenic) seeds mixed in with the Bt seeds being sold. Coined "Refuge-In-a-Bag" (RIB), this practice is intended to increase farmer compliance with refuge requirements and reduce additional labor needed at planting from having separate Bt and refuge seed bags on hand. This strategy is likely to reduce the likelihood of Bt-resistance occurring for corn rootworm, but may increase the risk of resistance for lepidopteran corn pests, such as European corn borer. Increased concerns for resistance with seed mixtures include partially resistant larvae on a Bt plant being able to move to a susceptible plant to survive or cross pollination of refuge pollen on to Bt plants that can lower the amount of Bt expressed in kernels for ear feeding insects.

Herbicide Resistance

Best management practices (BMPs) to control weeds may help delay resistance. BMPs include applying multiple herbicides with different modes of action, rotating crops, planting weed-free seed, scouting fields routinely, cleaning equipment to reduce the transmission of weeds to other fields, and maintaining field borders. The most widely planted GMOs are designed to tolerate herbicides. By 2006 some weed populations had evolved to tolerate some of the same herbicides. Palmer amaranth is a weed that competes with cotton. A native of the southwestern US, it traveled east and was first found resistant to glyphosate in 2006, less than 10 years after GM cotton was introduced.

Plant Protection

Farmers generally use less insecticide when they plant Bt-resistant crops. Insecticide use on corn farms declined from 0.21 pound per planted acre in 1995 to 0.02 pound in 2010. This is consistent with the decline in European corn borer populations as a direct result of Bt corn and cotton. The establishment of minimum refuge requirements helped delay the evolution of Bt resistance. However resistance appears to be developing to some Bt traits in some areas.

Tillage

By leaving at least 30% of crop residue on the soil surface from harvest through planting, conservation tillage reduces soil erosion from wind and water, increases water retention, and reduces soil degradation as well as water and chemical runoff. In addition, conservation tillage reduces the carbon footprint of agriculture. A 2014 review covering 12 states from 1996 to 2006, found that a 1% increase in herbicde-tolerant (HT) soybean adoption leads to a 0.21% increase in conservation tillage and a 0.3% decrease in quality-adjusted herbicide use.

Regulation

The regulation of genetic engineering concerns the approaches taken by governments to assess and manage the risks associated with the development and release of genetically modified crops. There are differences in the regulation of GM crops between countries, with some of the most marked differences occurring between the USA and Europe. Regulation varies in a given country depending on the intended use of each product. For example, a crop not intended for food use is generally not reviewed by authorities responsible for food safety.

Production

In 2013, GM crops were planted in 27 countries; 19 were developing countries and 8 were developed countries. 2013 was the second year in which developing countries grew a majority (54%) of the total GM harvest. 18 million farmers grew GM crops; around 90% were small-holding farmers in developing countries.

Country	2013– GM planted area (million hectares)	Biotech crops
USA	70.1	Maize, Soybean, Cotton, Canola, Sugarbeet, Alfalfa, Papaya, Squash
Brazil	40.3	Soybean, Maize, Cotton
Argentina	24.4	Soybean, Maize, Cotton
India	11.0	Cotton
Canada	10.8	Canola, Maize, Soybean, Sugarbeet
Total	175.2	----

The United States Department of Agriculture (USDA) reports every year on the total area of GMO varieties planted in the United States.According to National Agricultural Statistics Service, the states published in these tables represent 81–86 percent of all corn planted area, 88–90 percent of all soybean planted area, and 81–93 percent of all upland cotton planted area (depending on the year).

Global estimates are produced by the International Service for the Acquisition of Agri-biotech Applications (ISAAA) and can be found in their annual reports, "Global Status of Commercialized Transgenic Crops".

Farmers have widely adopted GM technology (see figure). Between 1996 and 2013, the total surface area of land cultivated with GM crops increased by a factor of 100, from 17,000 square kilometers (4,200,000 acres) to 1,750,000 km^2 (432 million acres). 10% of the world's arable land was planted with GM crops in 2010. As of 2011, 11 different transgenic crops were grown commercially on 395 million acres (160 million hectares) in 29 countries such as the USA, Brazil, Argentina, India, Canada, China, Paraguay, Pakistan, South Africa, Uruguay, Bolivia, Australia, Philippines, Myanmar, Burkina Faso, Mexico and Spain. One of the key reasons for this widespread adoption is the perceived economic benefit the technology brings to farmers. For example, the system of planting glyphosate-resistant seed and then applying glyphosate once plants emerged provided farmers with the opportunity to dramatically increase the yield from a given plot of land, since this allowed them to plant rows closer together. Without it, farmers had to plant rows far enough apart to control post-emergent weeds with mechanical tillage. Likewise, using Bt seeds means that farmers do not have to purchase insecticides, and then invest time, fuel, and equipment in applying them. However critics have disputed whether yields are higher and whether chemical use is less, with GM crops. See Genetically modified food controversies article for information.

In the US, by 2014, 94% of the planted area of soybeans, 96% of cotton and 93% of corn were genetically modified varieties. Genetically modified soybeans carried herbicide-tolerant traits only, but maize and cotton carried both herbicide tolerance and insect protection traits (the latter largely Bt protein). These constitute "input-traits" that are aimed to financially benefit the producers, but may have indirect environmental benefits and cost benefits to consumers. The Grocery Manufacturers of America estimated in 2003 that 70–75% of all processed foods in the U.S. contained a GM ingredient.

Europe grows relatively few genetically engineered crops with the exception of Spain, where one fifth of maize is genetically engineered, and smaller amounts in five other countries. The EU had a 'de facto' ban on the approval of new GM crops, from

1999 until 2004. GM crops are now regulated by the EU. In 2015, genetically engineered crops are banned in 38 countries worldwide, 19 of them in Europe. Developing countries grew 54 percent of genetically engineered crops in 2013.

In recent years GM crops expanded rapidly in developing countries. In 2013 approximately 18 million farmers grew 54% of worldwide GM crops in developing countries. 2013's largest increase was in Brazil (403,000 km^2versus 368,000 km^2 in 2012). GM cotton began growing in India in 2002, reaching 110,000 km^2 in 2013.

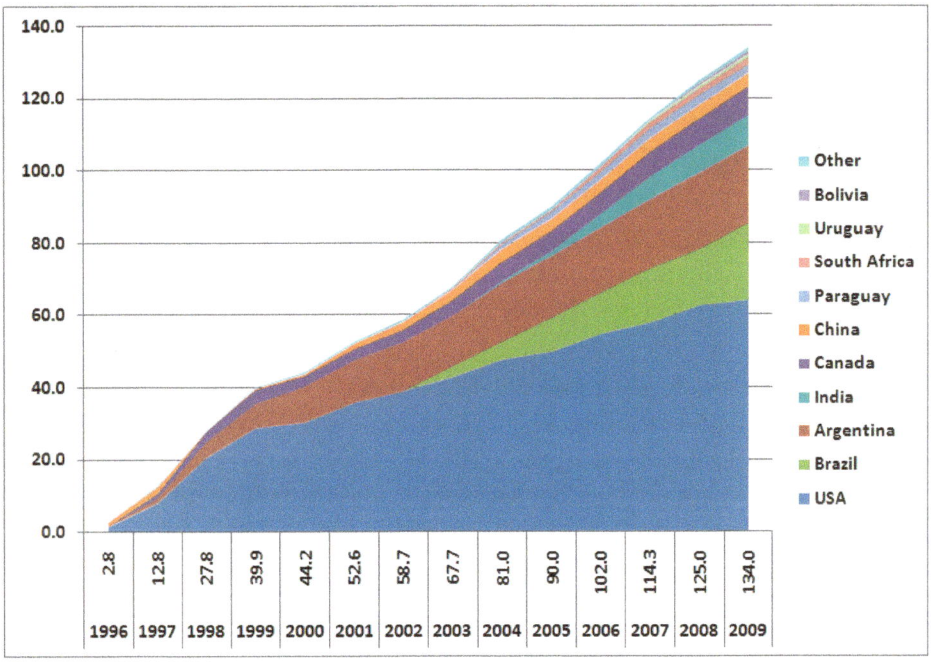

Land area used for genetically modified crops by country (1996–2009), in millions of hectares. In 2011, the land area used was 160 million hectares, or 1.6 million square kilometers.

According to the 2013 ISAAA brief: "...a total of 36 countries (35 + EU-28) have granted regulatory approvals for biotech crops for food and/or feed use and for environmental release or planting since 1994... a total of 2,833 regulatory approvals involving 27 GM crops and 336 GM events (NB: an "event" is a specific genetic modification in a specific species) have been issued by authorities, of which 1,321 are for food use (direct use or processing), 918 for feed use (direct use or processing) and 599 for environmental release or planting. Japan has the largest number (198), followed by the U.S.A. (165, not including "stacked" events), Canada (146), Mexico (131), South Korea (103), Australia (93), New Zealand (83), European Union (71 including approvals that have expired or under renewal process), Philippines (68), Taiwan (65), Colombia (59), China (55) and South Africa (52). Maize has the largest number (130 events in 27 countries), followed by cotton (49 events in 22 countries), potato (31 events in 10 countries), canola (30 events in 12 countries) and soybean (27 events in 26 countries).

Controversy

GM foods are controversial and the subject of protests, vandalism, referenda, legislation, court action and scientific disputes. The controversies involve consumers, biotechnology companies, governmental regulators, non-governmental organizations and scientists. The key areas are whether GM food should be labeled, the role of government regulators, the effect of GM crops on health and the environment, the effects of pesticide use and resistance, the impact on farmers, and their roles in feeding the world and energy production.

There is a scientific consensus that currently available food derived from GM crops poses no greater risk to human health than conventional food, but that each GM food needs to be tested on a case-by-case basis before introduction.Nonetheless, members of the public are much less likely than scientists to perceive GM foods as safe. The legal and regulatory status of GM foods varies by country, with some nations banning or restricting them, and others permitting them with widely differing degrees of regulation.

No reports of ill effects have been documented in the human population from GM food. Although GMO labeling is required in many countries, the United States Food and Drug Administration does not require labeling, nor does it recognize a distinction between approved GMO and non-GMO foods.

Advocacy groups such as Center for Food Safety, Union of Concerned Scientists, Greenpeace and the World Wildlife Fund claim that risks related to GM food have not been adequately examined and managed, that GMOs are not sufficiently tested and should be labelled, and that regulatory authorities and scientific bodies are too closely tied to industry. Some studies have claimed that genetically modified crops can cause harm;a 2016 review that reanalyzed the data from six of these studies found that their statistical methodologies were flawed and did not demonstrate harm, and said that conclusions about GMO crop safety should be drawn from "the totality of the evidence... instead of far-fetched evidence from single studies".

REFERENCES

1. ISAAA 2013 Annual Report Executive Summary, Global Status of Commercialized Biotech/GM Crops: 2013 ISAAA Brief 46-2013, Retrieved 6 August 2014
2. ISAAA 2015 Annual Report Executive Summary, 20th Anniversary (1996 to 2015) of the Global Commercialization of Biotech Crops and Biotech Crop Highlights in 2015 ISAAA Brief 51-2015, Retrieved 19 August 2016
3. James, C (2011). "ISAAA Brief 43, Global Status of Commercialized Biotech/GM Crops: 2011". ISAAA Briefs. Ithaca, New York: International Service for the Acquisition of Agri-biotech Applications (ISAAA). Retrieved 2012-06-02.
4. Adoption of Genetically Engineered Crops in the U.S. – Economic Research Service, of the U.S. Department of Agriculture
5. Klümper, W; Qaim, M (2014). "A Meta-Analysis of the Impacts of Genetically Modified Crops". PLoS ONE. 9 (11): e111629. Bibcode:2014PLoSO...9k1629K. doi:10.1371/journal.pone.0111629. PMC 4218791 . PMID 25365303.

6. Andrew Pollack for the New York Times. April 13, 2010 Study Says Overuse Threatens Gains From Modified Crops
7. Nicolia, Alessandro; Manzo, Alberto; Veronesi, Fabio; Rosellini, Daniele (2013). "An overview of the last 10 years of genetically engineered crop safety research" (PDF). Critical Reviews in Biotechnology. 34 (1): 1–12. doi:10.3109/07388551.2013.823595. PMID 24041244.
8. "State of Food and Agriculture 2003–2004. Agricultural Biotechnology: Meeting the Needs of the Poor. Health and environmental impacts of transgenic crops".
9. Ronald, Pamela (May 5, 2011). "Plant Genetics, Sustainable Agriculture and Global Food Security". Genetics. 188 (1): 11–20. doi:10.1534/genetics.111.128553. PMC 3120150 . PMID 21546547.
10. Domingo, José L.; Bordonaba, Jordi Giné (2011). "A literature review on the safety assessment of genetically modified plants" (PDF). Environment International. 37 (4): 734–742. doi:10.1016/j.envint.2011.01.003. PMID 21296423. .
11. "Statement by the AAAS Board of Directors On Labeling of Genetically Modified Foods" (PDF).
12. "A decade of EU-funded GMO research (2001–2010)" (PDF). Directorate-General for Research and Innovation. Biotechnologies, Agriculture, Food. European Commission, European Union. 2010. doi:10.2777/97784. ISBN 978-92-79-16344-9. Retrieved February 8, 2016.
13. "AMA Report on Genetically Modified Crops and Foods (online summary)".
14. "REPORT 2 OF THE COUNCIL ON SCIENCE AND PUBLIC HEALTH (A-12): Labeling of Bioengineered Foods" .
15. "Restrictions on Genetically Modified Organisms: United States. Public and Scholarly Opinion".
16. "Genetically Engineered Crops: Experiences and Prospects".
17. "Frequently asked questions on genetically modified foods"..
18. Haslberger, Alexander G. (2003). "Codex guidelines for GM foods include the analysis of unintended effects". Nature Biotechnolgy. 21 (7): 739–741. doi:10.1038/nbt0703-739. PMID 12833088. These principles dictate a case-by-case premarket assessment that includes an evaluation of both direct and unintended effects.
19. Some medical organizations, including the British Medical Association, advocate further caution based upon the precautionary principle:
20. Funk, Cary; Rainie, Lee (January 29, 2015). "Public and Scientists' Views on Science and Society". Pew Research Center. Retrieved February 24, 2016. The largest differences between the public and the AAAS scientists are found in beliefs about the safety of eating genetically modified (GM) foods. Nearly nine-in-ten (88%) scientists say it is generally safe to eat GM foods compared with 37% of the general public, a difference of 51 percentage points.
21. Marris, Claire (2001). "Public views on GMOs: deconstructing the myths" (PDF). EMBO Reports. 2 (7): 545–548. doi:10.1093/embo-reports/kve142. PMC

1083956 . PMID 11463731.

22. Final Report of the PABE research project (December 2001). "Public Perceptions of Agricultural Biotechnologies in Europe". Commission of European Communities. Retrieved February 24, 2016.
23. Scott, Sydney E.; Inbar, Yoel; Rozin, Paul (2016). "Evidence for Absolute Moral Opposition to Genetically Modified Food in the United States" (PDF). Perspectives on Psychological Science. 11 (3): 315–324. doi:10.1177/1745691615621275. PMID 27217243.
24. "Restrictions on Genetically Modified Organisms". Library of Congress. June 9, 2015. Retrieved February 24, 2016.
25. Bashshur, Ramona (February 2013). "FDA and Regulation of GMOs". American Bar Association. Retrieved February 24, 2016.
26. Sifferlin, Alexandra (October 3, 2015). "Over Half of E.U. Countries Are Opting Out of GMOs". Time.
27. Lynch, Diahanna; Vogel, David (April 5, 2001). "The Regulation of GMOs in Europe and the United States: A Case-Study of Contemporary European Regulatory Politics". Council on Foreign Relations. Retrieved February 24, 2016.
28. Lederberg J, Tatum EL; Tatum (1946). "Gene recombination in E. coli". Nature. 158 (4016): 558. Bibcode:1946Natur.158..558L. doi:10.1038/158558a0.
29. Bock, R. (2010). "The give-and-take of DNA: horizontal gene transfer in plants". Trends in Plant Science. 15 (1): 11–22. doi:10.1016/j.tplants.2009.10.001. PMID 19910236.
30. Morgante, M.; Brunner, S.; Pea, G.; Fengler, K.; Zuccolo, A.; Rafalski, A. (2005). "Gene duplication and exon shuffling by helitron-like transposons generate intraspecies diversity in maize". Nature Genetics. 37 (9): 997–1002. doi:10.1038/ng1615. PMID 16056225.
31. Monroe D. (2006). "Jumping Genes Cross Plant Species Boundaries". PLoS Biology. 4 (1): e35. doi:10.1371/journal.pbio.0040035.
32. Feschotte, C.; Osterlund, M. T.; Peeler, R.; Wessler, S. R. (2005). "DNA-binding specificity of rice mariner-like transposases and interactions with Stowaway MITEs". Nucleic Acids Research. 33 (7): 2153–65. doi:10.1093/nar/gki509. PMC 1079968 . PMID 15831788.
33. Cordaux, R.; Udit, S.; Batzer, M.; Feschotte, C. (2006). "Birth of a chimeric primate gene by capture of the transposase gene from a mobile element". Proceedings of the National Academy of Sciences of the United States of America. 103 (21): 8101–8106. Bibcode:2006PNAS..103.8101C. doi:10.1073/pnas.0601161103. PMC 1472436 . PMID 16672366.
34. Long, M.; Betrán, E.; Thornton, K.; Wang, W. (2003). "The origin of new genes: glimpses from the young and old". Nature Reviews Genetics. 4 (11): 865–875. doi:10.1038/nrg1204. PMID 14634634.
35. Chen, Z. (2010). "Molecular mechanisms of polyploidy and hybrid vigor". Trends in Plant Science. 15 (2): 57–71. doi:10.1016/j.tplants.2009.12.003. PMC

2821985 . PMID 20080432.

36. Hoisington D, Khairallah M, Reeves T, Ribaut JM, Skovmand B, Taba S, Warburton M; Khairallah; Reeves; Ribaut; Skovmand; Taba; Warburton (1999). "Plant genetic resources: What can they contribute toward increased crop productivity?" (PDF). Proceedings of the National Academy of Sciences of the United States of America. 96 (11): 5937–43. Bibcode:1999PNAS...96.5937H. doi:10.1073/pnas.96.11.5937. ISSN 0027-8424. PMC 34209 . PMID 10339521.
37. Predieri, S. (2001). "Mutation induction and tissue culture in improving fruits". Plant Cell, Tissue and Organ Culture. 64 (2/3): 185–210. doi:10.1023/A:1010623203554.
38. Duncan, R. (1996). "Tissue Culture-Induced Variation and Crop Improvement". Advances in Agronomy. Advances in Agronomy. 58: 201–240. doi:10.1016/S0065-2113(08)60256-4. ISBN 9780120007585.
39. Fraley, RT; et al. (1983). "Expression of bacterial genes in plant cells" (PDF). Proc. Natl. Acad. Sci. U.S.A. 80: 4803–4807. Bibcode:1983PNAS...80.4803F.
40. James, Clive (1996). "Global Review of the Field Testing and Commercialization of Transgenic Plants: 1986 to 1995" (PDF). The International Service for the Acquisition of Agri-biotech Applications. Retrieved 17 July 2010.
41. Vaeck, M; et al. (1987). "Transgenic plants protected from insect attack" (PDF). Nature. 328 (6125): 33–37. Bibcode:1987Natur.328...33V. doi:10.1038/328033a0.
42. James, C (1997). "Global Status of Transgenic Crops in 1997"(PDF). ISAAA Briefs No. 5.: 31.
43. Conner, AJ; Glare, TR; Nap, JP (January 2003). "The release of genetically modified crops into the environment. Part II. Overview of ecological risk assessment". Plant J. 33 (1): 19–46. doi:10.1046/j.0960-7412.2002.001607.x. PMID 12943539.
44. Bruening, G.; Lyons, J. M. (2000). "The case of the FLAVR SAVR tomato". California Agriculture. 54 (4): 6–7. doi:10.3733/ca.v054n04p6.
45. Debora MacKenzie (18 June 1994). "Transgenic tobacco is European first". New Scientist.
46. Genetically Altered Potato Ok'd For Crops Lawrence Journal-World – 6 May 1995
47. Andrew Pollack (19 June 2013). "Executive at Monsanto wins global food honor". The New York Times. Retrieved 20 June 2013.
48. Rebecca Boyle for Popular Science. January 24, 2011. How To Genetically Modify a Seed, Step By Step
49. "Bombarded - Define Bombarded at Dictionary.com". Dictionary.com.
50. Shrawat, A.; Lörz, H. (2006). "Agrobacterium-mediated transformation of cereals: a promising approach crossing barriers". Plant biotechnology journal. 4 (6): 575–603. doi:10.1111/j.1467-7652.2006.00209.x. PMID 17309731.
51. Maghari, Behrokh Mohajer, and Ali M. Ardekani. "Genetically Modified Foods And Social Concerns." Avicenna Journal Of Medical Biotechnology 3.3 (2011): 109-117. Academic Search Premier. Web. 7 Nov. 2014.

52. "Information Systems for Biotechnology News Report".
53. Catchpole, G. S.; Beckmann, M.; Enot, D. P.; Mondhe, M.; Zywicki, B.; Taylor, J.; Hardy, N.; Smith, A.; King, R. D.; Kell, D. B.; Fiehn, O.; Draper, J. (2005). "Hierarchical metabolomics demonstrates substantial compositional similarity between genetically modified and conventional potato crops". Proceedings of the National Academy of Sciences. 102 (40): 14458–14462. Bibcode:2005PNAS..10214458C. doi:10.1073/pnas.0503955102.
54. Koornneef, M.; Meinke, D. (2010). "The development of Arabidopsis as a model plant". The Plant journal : for cell and molecular biology. 61 (6): 909–921. doi:10.1111/j.1365-313X.2009.04086.x. PMID 20409266.
55. Banjara, Manoj; Zhu, Longfu; Shen, Guoxin; Payton, Paxton; Zhang, Hong (2012-01-01). "Expression of an Arabidopsis sodium/proton antiporter gene (AtNHX1) in peanut to improve salt tolerance". Plant Biotechnology Reports. 6: 59–67. doi:10.1007/s11816-011-0200-5. Retrieved 2013-12-19.
56. Robin McKie. "GM corn set to stop man spreading his seed". the Guardian.
57. Walmsley, A.; Arntzen, C. (2000). "Plants for delivery of edible vaccines". Current Opinion in Biotechnology. 11 (2): 126–9. doi:10.1016/S0958-1669(00)00070-7. PMID 10753769.
58. Nancy Podevina & Patrick du Jardin (2012). "Possible consequences of the overlap between the CaMV 35S promoter regions in plant transformation vectors used and the viral gene VI in transgenic plants". GM Crops & Food: Biotechnology in Agriculture and the Food Chain. 3 (4): 296–300. doi:10.4161/gmcr.21406. PMID 22892689.
59. Maxmen, Amy (2 May 2012) First plant-made drug on the marketNature, Biology & Biotechnology, Industry. Retrieved 26 June 2012
60. NWT magazine, April 2011
61. "Molecular Physiology".
62. Project by Dean Price increasing photosynthesis by 15 to 25%
63. Additional project by Dean Price; adding of CO^2-concentrating cage
64. "Brocade Desktop: irua".
65. "Auxin patterns Solanum lycopersicum leaf morphogenesis".
66. Projects changing respectively plant growth and plant flowers
67. Project changing number of stomata in plants conducted by Ikuko Hara-Nishimura
68. (4 May 2013) One Per Cent: Grow your own living lights The New Scientist, Issue 2915, Retrieved 7 May 2013
69. MacKenzie, Deborah (2 August 2008). "How the humble potato could feed the world" (cover story) New Scientist. No 2667 pp.30–33
70. Talbot, David (2014-07-19). "Beijing Researchers Use Gene Editing to Create Disease-Resistant Wheat | MIT Technology Review". Technologyreview.com. Retrieved 2014-07-23.
71. Wang, Yanpeng (2014). "Simultaneous editing of three homoeoalleles in

hexaploid bread wheat confers heritable resistance to powdery mildew". Nature Biotechnology. 32 (9): 947–951. doi:10.1038/nbt.2969.

72. Waltz, Emily. "Gene-edited CRISPR mushroom escapes US regulation". nature.com. Nature. Retrieved 18 April 2016.
73. Economic Impact of Transgenic Crops in Developing Countries. Agbioworld.org. Retrieved 8 February 2011.
74. Areal, F. J.; Riesgo, L.; Rodríguez-Cerezo, E. (2012). "Economic and agronomic impact of commercialized GM crops: A meta-analysis". The Journal of Agricultural Science. 151: 7–33. doi:10.1017/S0021859612000111.
75. Finger, Robert; El Benni, Nadja; Kaphengst, Timo; Evans, Clive; Herbert, Sophie; Lehmann, Bernard; Morse, Stephen; Stupak, Nataliya (2011). "A Meta Analysis on Farm-Level Costs and Benefits of GM Crops". Sustainability. 3 (12): 743–762. doi:10.3390/su3050743.
76. Hutchison WD, Burkness EC, Mitchell PD, Moon RD, Leslie TW, Fleischer SJ, Abrahamson M, Hamilton KL, Steffey KL, Gray ME, Hellmich RL, Kaster LV, Hunt TE, Wright RJ, Pecinovsky K, Rabaey TL, Flood BR, Raun ES (October 2010). "Areawide suppression of European corn borer with Bt maize reaps savings to non-Bt maize growers". Science. 330 (6001): 222–5. Bibcode:2010Sci...330..222H. doi:10.1126/science.1190242. PMID 20929774.
77. Karnowski, Steve High-Tech Corn Fights Pests at Home and NearbySci-Tech today, 8 October 2010. Retrieved 9 October 2010.
78. Falck-Zepeda, José Benjamin; Traxler, Greg; Nelson, Robert G. (2000). "Surplus Distribution from the Introduction of a Biotechnology Innovation". American Journal of Agricultural Economics. 82 (2): 360–9. doi:10.1111/0002-9092.00031. JSTOR 1244657.
79. James, C. 2014 Global Status of Commercialized Biotech/GM Crops: 2014 ISAAA Brief No. 49.
80. Brookes, Graham and Barfoot, Peter (May 2012) GM crops: global socio-economic and environmental impacts 1996-2010 PG Economics Ltd. UK, Retrieved 3 January 2012
81. Smale, M., P. Zambrano, and M. Cartel (2006). "Bales and balance: A review of the methods used to assess the economic impact of Bt cotton on farmers in developing economies" (PDF). AgBioForum. 9 (3): 195–212.
82. Planting the future: opportunities and challenges for using crop genetic improvement technologies for sustainable agriculture, EASAC policy report 21, 27.06.13.
83. Rajib Deb, et al. "Genetically Modified (Gm) Crops Lifeline For Livestock – A Review." Agricultural Reviews 31.4 (2010): 279–285.
84. Robert Langreth and Matthew Herper for Forbes. December 31, 2009 The Planet Versus Monsanto
85. Cavallaro, Matt (June 26, 2009). "The Seeds Of A Monsanto Short Play". Forbes.
86. Regalado, Antonio (July 30, 2015). "Monsanto Roundup Ready Soybean Patent

Expiration Ushers in Generic GMOs | MIT Technology Review". MIT Technology Review. Retrieved 2015-10-22.

87. "Monsanto Will Let Bio-Crop Patents Expire". Business Week. January 21, 2010.
88. Monsanto. Roundup Ready Soybean Patent Expiration
89. "Monsanto ~ Licensing". Monsanto.com. November 3, 2008.
90. Monsanto GMO Ignites Big Seed War. NPR.
91. "Syngenta US | Corn and Soybean Seed – Garst, Golden Harvest, NK, Agrisure". Syngenta.com.
92. "Agronomy Library – Pioneer Hi-Bred Agronomy Library". Pioneer.com.
93. "Genetically modified crops - Field research". Economist. November 8, 2014. Retrieved October 3, 2016.
94. "SeedQuest - Central information website for the global seed industry".
95. ISAAA Pocket K No. 35: Bt Brinjal in India
96. Andrew Pollack for the New York Times. 7 November 2014. U.S.D.A. Approves Modified Potato. Next Up: French Fry Fans
97. Federal Register. May 3, 2013. J.R. Simplot Co.; Availability of Petition for Determination of Nonregulated Status of Potato Genetically Engineered for Low Acrylamide Potential and Reduced Black Spot Bruise
98. Pollack, A. "Gene-Altered Apples Get U.S. Approval" New York Times. February 13, 2015.
99. Tennille, Tracy (February 13, 2015). "First Genetically Modified Apple Approved for Sale in U.S.". Wall Street Journal. Retrieved October 3, 2016.
100. "Apple-to-apple transformation" Archived 25 September 2013 at the Wayback Machine.. Okanagan Specialty Fruits. Retrieved 2012-08-03.
101. "Artic apples FAQ". Arctic Apples. 2014. Retrieved October 3, 2016.
102. "FDA concludes Arctic Apples and Innate Potatoes are safe for consumption". United States Food and Drug Administration. 20 March 2015.
103. Canadian Food Inspection Agency. DD2009-76: Determination of the Safety of Pioneer Hi-Bred Production Ltd.'s Soybean (Glycine max (L.) Merr.) Event 305423 Issued: 2009-04 . Retrieved January 2011
104. Andrew Pollack for the New York Times. November 15, 2013 In a Bean, a Boon to Biotech
105. Crop plants – "green factories" for fish oils, Rothamsted Research 14-11-2013.
106. Ruiz-Lopez, Noemi; et al. (2013). "Successful high-level accumulation of fish oil omega-3 long chain polyunsaturated fatty acids in a transgenic oilseed crop". The Plant Journal. 77 (2): 198–208. doi:10.1111/tpj.12378. PMC 4253037 . PMID 24308505.
107. About Golden Rice Archived 2 November 2012 at the Wayback Machine. International Rice Research Institute. Retrieved 20 August 2012
108. Nayar, A. (2011). "Grants aim to fight malnutrition". Nature. doi:10.1038/

news.2011.233.

109. Philpott, Tom (3 February 2016). "WTF Happened to Golden Rice?". Mother Jones. Retrieved 24 March 2016.

110. "Transgenic multivitamin corn through biofortification of endosperm with three vitamins representing three distinct metabolic pathways".

111. BROWNSTONE, SYDNEY (June 30, 2014). "Humans Are About To Taste The First Genetically Engineered "Super" Bananas". Co.exist.

112. Sayre, R.; Beeching, J. R.; Cahoon, E. B.; Egesi, C.; Fauquet, C.; Fellman, J.; Fregene, M.; Gruissem, W.; Mallowa, S.; Manary, M.; Maziya-Dixon, B.; Mbanaso, A.; Schachtman, D. P.; Siritunga, D.; Taylor, N.; Vanderschuren, H.; Zhang, P. (2011). "The BioCassava Plus Program: Biofortification of Cassava for Sub-Saharan Africa". Annual Review of Plant Biology. 62: 251–272. doi:10.1146/annurev-arplant-042110-103751. PMID 21526968.

113. Paarlburg, Robert Drought Tolerant GMO Maize in Africa, Anticipating Regulatory Hurdles Archived 22 December 2014 at the Wayback Machine. International Life Sciences Institute, January 2011. Retrieved 25 April 2011

114. Australia continues to test drought-resistant GM wheat GMO Compass, 16 July 2008. Retrieved 25 April 2011

115. Staff (14 May 2011) USA: USDA allows large-scale GM eucalyptus trial GMO Cmpass. Retrieved 29 September 2011

116. Lundmark, C. (2006). "Searching Evolutionary Pathways: Antifreeze Genes from Antarctic Hairgrass". BioScience. 56 (6): 552. doi:10.1641/0006-3568(2006)56552:SEP2.0.CO;2. ISSN 0006-3568.

117. Banjara, Manoj; Longfu Zhu; Guoxin Shen; Paxton Payton; Hong Zhang (2012). "Expression of an Arabidopsis sodium/proton antiporter gene (AtNHX1) in peanut to improve salt tolerance". Plant Biotechnol Rep. 6: 59–67. doi:10.1007/s11816-011-0200-5.

118. Sara Abdulla (27 May 1999). "Drought stress". Nature News. doi:10.1038/news990527-9 (inactive 2017-01-16).

119. Rennie, Rob and Heffer, Patrick Anticipated Impact of Modern Biotechnology on Nutrient Use Efficiency TFI/FIRT Fertilizer Outlook and Technology Conference 16–18 November 2010, Savannah (GA), Web page. Retrieved 25 April 2011

120. Michael Eisenstein Plant breeding: Discovery in a dry spell Nature 501, S7–S9 (26 September 2013) Published online 25 September 2013

121. Carpenter, J.; Gianessi, L. (1999). "Herbicide tolerant soybeans: Why growers are adopting Roundup Ready varieties". AgBioForum. 2 (2): 65–72.

122. G. R. Heck; et al. (1 January 2005). "Development and Characterization of a CP4 EPSPS-Based, Glyphosate-Tolerant Corn Event" (Free full text). Crop Sci. 45 (1): 329–339. doi:10.2135/cropsci2005.0329.

123. T. Funke; et al. (2006). "Molecular basis for the herbicide resistance of Roundup Ready crops" (Free full text). PNAS. 103 (35): 13010–13015. Bibcode:2006PNAS..10313010F. doi:10.1073/pnas.0603638103. PMC 1559744

. PMID 16916934.

124. L. P. Gianessi, C. S. Silvers, S. Sankula and J. E. Carpenter. Plant Biotechnology: Current and Potential Impact for Improving Pest management in US Agriculture, An Analysis of 40 Case Studies(Washington, D.C.: National Center for Food and Agricultural Policy, 2002), 5–6
125. Kasey J. (8 September 2011)Attack of the Superweed. Bloomberg Businessweek.
126. Mark Ganchiff for Midwest Wine Press. August 24, 2013 New Herbicide Resistant Crops Being Considered By USDA
127. ISAAA GM Approval Database Gene list: aad1. Page accessed February 27, 2015
128. EPA Press Release. October 15, 2014 EPA Announces Final Decision to Register Enlist Duo, Herbicide Containing 2, 4-D and Glyphosate/Risk assessment ensures protection of human health, including infants, children EPA Documents: Registration of Enlist Duo
129. Mark A. Peterson et al. for ISB News Report, May 2011 Utility of Aryloxyalkanoate Dioxygenase Transgenes for Development of New Herbicide Resistant Crop Technologies
130. Colin Schultz for The Smithsonian.com. September 25, 2014 The USDA Approved a New GM Crop to Deal With Problems Created by the Old GM Crops
131. Johnson, William G.; Hallett, Steven G.; Legleiter, Travis R.; Whitford, Fred; Weller, Stephen C.; Bordelon, Bruce P.; Lerner, B. Rosie (November 2012). "2,4-D- and Dicamba-tolerant Crops—Some Facts to Consider" (PDF). Purdue University Extension. Retrieved October 3, 2016.
132. Vaeck, M.; Reynaerts, A.; Höfte, H.; Jansens, S.; De Beuckeleer, M.; Dean, C.; Zabeau, M.; Montagu, M. V.; Leemans, J. (1987). "Transgenic plants protected from insect attack". Nature. 328 (6125): 33–37. Bibcode:1987Natur.328...33V. doi:10.1038/328033a0.
133. National Academy of Sciences (2001). Transgenic Plants and World Agriculture. Washington: National Academy Press.
134. Naranjo, Steven (April 22, 2008). "The Present and Future Role of Insect-Resistant Genetically Modified Cotton in IPM" (PDF). USDA.gov. United States department of agriculture. Retrieved December 3, 2015.
135. Erica Kipp for the Botany Global Issues Map. February, 2000. Genetically Altered Papayas Save the Harvest Archive date December 13, 2004.
136. "The Rainbow Papaya Story". Hawaii Papaya Industry Association. 2006. Retrieved 2014-12-27.
137. Ronald, Pamela and McWilliams, James for the New York Times, 14 May 2010. Genetically Engineered Distortions
138. Wenslaff, Timothy F.; Osgood, Robert V. (October 2000). "Production Of UH Sunup Transgenic Papaya Seed In Hawaii" (PDF). Hawaii Agriculture Research Center. Archived from the original (PDF) on 31 March 2012.
139. "Genetically Engineered Foods - Plant Virus Resistance" (PDF). Cornell Cooperative Extension. Cornell University. 2002. Retrieved October 3, 2016.

140. "How Many Foods Are Genetically Engineered?". University of California. 2012-02-16. Retrieved October 3, 2016.
141. Wang, Guo-ying (2009). "Genetic Engineering for Maize Improvement in China". Electronic Journal of Biotechnology. Electronic Journal of Biotechnology. Retrieved December 1, 2015.
142. Gali Weinreb and Koby Yeshayahou for Globes May 2, 2012. FDA approves Protalix Gaucher treatment Archived 29 May 2013 at the Wayback Machine.
143. Jha, Alok (14 August 2012) Julian Ma: I'm growing antibodies in tobacco plants to help prevent HIV The Guardian. Retrieved 12 March 2012
144. Carrington, Damien (19 January 2012) GM microbe breakthrough paves way for large-scale seaweed farming for biofuels The Guardian. Retrieved 12 March 2012
145. http://www.isaaa.org/kc/cropbiotechupdate/article/default.asp?ID=9597
146. Carolyn Lochhead for the San Francisco Chronicle. 30 April 2012 Genetically modified crops' results raise concern
147. Hope, Alan (3 April 2013, News in brief: The Bio Safety Council..."Flanders Today, Page 2, Retrieved 27 April 2013
148. (2013) Wout Boerjan Lab VIB (Flemish Institute for Biotechnology) Gent, Retrieved 27 April 2013
149. van Beilen, Jan B.; Yves Poirier (May 2008). "Harnessing plant biomass for biofuels and biomaterials:Production of renewable polymers from crop plants". The Plant Journal. 54 (4): 684–701. doi:10.1111/j.1365-313X.2008.03431.x. PMID 18476872.
150. "The History and Future of GM Potatoes". PotatoPro Newsletter. March 10, 2010.
151. Strange, Amy (20 September 2011) Scientists engineer plants to eat toxic pollution The Irish Times. Retrieved 20 September 2011
152. Chard, Abigail (2011) Growing a grass that loves bombs The British Science Association. Retrieved 20 September 2011 Archived24 July 2012 at the Wayback Machine.
153. Langston, Jennifer (2016-11-22). "New grasses neutralize toxic pollution from bombs, explosives, and munitions". ScienceDaily. Retrieved 2016-11-30.
154. Meagher, RB (2000). "Phytoremediation of toxic elemental and organic pollutants". Current Opinion in Plant Biology. 3 (2): 153–162. doi:10.1016/S1369-5266(99)00054-0. PMID 10712958.
155. Martins VAP (2008). "Genomic Insights into Oil Biodegradation in Marine Systems". Microbial Biodegradation: Genomics and Molecular Biology. Caister Academic Press. ISBN 978-1-904455-17-2.
156. Daniel Charles (1 March 2003). "Corn That Clones Itself". Technology Review.
157. "GM Crops List | GM Approval Database- ISAAA.org". www.isaaa.org. Retrieved 2016-01-30.
158. "All the GMOs Approved In the U.S.". TIME.com. Retrieved 2016-02-11.
159. www.gmo-compass.org. "Lucerne - GMO Database". www.gmo-compass.org.

Retrieved 2016-02-11.

160. "UPDATE 3-U.S. farmers get approval to plant GMO alfalfa". Reuters. 2011-01-27. Retrieved 2016-02-11.
161. "Infographics: Global Status of Commercialized Biotech/GM Crops: 2014 - ISAAA Brief 49-2014 | ISAAA.org". www.isaaa.org. Retrieved 2016-02-11.
162. Scott Kilman. "Modified Beet Gets New Life". Wall Street Journal. Retrieved 2016-02-15.
163. Pollack, Andrew (2007-11-27). "Round 2 for Biotech Beets". The New York Times. ISSN 0362-4331. Retrieved 2016-02-15.
164. "Facts and trends - India" (PDF). International Service for the Acquisition of Agri-biotech Applications.
165. "Executive Summary: Global Status of Commercialized Biotech/GM Crops: 2014 - ISAAA Brief 49-2014 | ISAAA.org". www.isaaa.org. Retrieved 2016-02-16.
166. "Facts and trends-Mexico" (PDF). International Service for the Acquisition of Agri-biotech Applications.
167. "Facts and trends- China" (PDF). International Service for the Acquisition of Agri-biotech Applications.
168. "Facts and trends - Columbia" (PDF). International Service for the Acquisition of Agri-biotech Applications.
169. Carter, Colin; Moschini, GianCarlo; Sheldon, Ian, eds. (2011). Genetically Modified Food and Global Welfare (Frontiers of Economics and Globalization). United Kingdom: Emerald Group Publishing Limited. p. 89. ISBN 978-0857247575.
170. Press, Associated (2010-03-03). "GM potato to be grown in Europe". The Guardian. ISSN 0261-3077. Retrieved 2016-02-15.
171. Fernandez-Cornejo, Jorge; Wechsler, Seth; Livingston, Mike; Mitchell, Lorraine (February 2014). "Genetically Engineered Crops in the United States (summary)" (PDF). Economic Research Service USDA. United States Department of Agriculture. p. 2. Retrieved October 3, 2016.
172. Tabashnik; et al. (2003). "Insect Resistance to Transgenic Bt Crops: Lessons from the Laboratory and Field" (PDF). J. Econ. Entomol. 96 (4): 1031–1038. doi:10.1603/0022-0493-96.4.1031. PMID 14503572.
173. Roush RT (1997). "Bt-transgenic crops: just another pretty insecticide or a chance for a new start in the resistance management?". Pestic. Sci. 51 (3): 328–34. doi:10.1002/(SICI)1096-9063(199711)51:3<328::AID-PS650>3.0.CO;2-B.
174. Dong, H. Z.; Li, W. J. (2007). "Variability of Endotoxin Expression in Bt Transgenic Cotton". Journal of Agronomy & Crop Science. 193: 21–9. doi:10.1111/j.1439-037X.2006.00240.x.
175. Tabashnik BE, Carrière Y, Dennehy TJ; Carrière; Dennehy; Morin; Sisterson; Roush; Shelton; Zhao (August 2003). "Insect resistance to transgenic Bt crops: lessons from the laboratory and field". J. Econ. Entomol. 96 (4): 1031–8. doi:10.1603/0022-0493-96.4.1031. PMID 14503572.

176. APPDMZ\ccvivr. "Monsanto - Pink Bollworm Resistance to GM Cotton in India".
177. "The Real Deal: Explaining Monsanto's Refuge-in-the-Bag Concept". www.monsanto.com. Retrieved 2015-12-03.
178. Siegfried, B.D.; et al. (2012). "Understanding successful resistance management". GM Crops & Food. 3 (3): 184–193. doi:10.4161/gmcr.20715.
179. Devos, Y.; et al. (2013). "Resistance evolution to the first generation of genetically modified Diabrotica-active Bt-maize events by western corn rootworm: management and monitoring considerations". Transgenic Research. 22 (2): 269–299. doi:10.1007/s11248-012-9657-4. PMID 23011587.
180. Culpepper, Stanley A; et al. (2006). "Glyphosate-resistant Palmer amaranth (Amaranthus palmeri) confirmed in Georgia". Weed Science. 54 (4): 620–626. doi:10.1614/ws-06-001r.1.
181. Gallant, Andre. "Pigweed in the Cotton: A superweed invades Georgia". Modern Farmer.
182. Webster, TM; Grey, TL (2015). "Glyphosate-Resistant Palmer Amaranth (Amaranthus palmeri) Morphology, Growth, and Seed Production in Georgia.". Weed Science. 63 (1): 264–272. doi:10.1614/ws-d-14-00051.1.
183. Fernandez-Cornejo, Jorge; Hallahan,, Charlie; Nehring,, Richard; Wechsler, Seth; Grube, Arthur (2014). "Conservation Tillage, Herbicide Use, and Genetically Engineered Crops in the United States: The Case of Soybeans". AgBioForum. 15 (3). Retrieved October 3, 2016.
184. Wesseler, J. and N. Kalaitzandonakes (2011): Present and Future EU GMO policy. In Arie Oskam, Gerrit Meesters and Huib Silvis (eds.), EU Policy for Agriculture, Food and Rural Areas. Second Edition, pp. 23-323 – 23-332. Wageningen: Wageningen Academic Publishers
185. Beckmann, V., C. Soregaroli, J. Wesseler (2011): Coexistence of genetically modified (GM) and non-modified (non GM) crops: Are the two main property rights regimes equivalent with respect to the coexistence value? In "Genetically modified food and global welfare" edited by Colin Carter, GianCarlo Moschini and Ian Sheldon, pp 201-224. Volume 10 in Frontiers of Economics and Globalization Series. Bingley, UK: Emerald Group Publishing
186. ISAAA 2012 Annual Report Executive Summary
187. Fernandez-Cornejo, Jorge (1 July 2009). Adoption of Genetically Engineered Crops in the U.S. Data Sets. Economic Research Service, United States Department of Agriculture. OCLC 53942168. Archived from the original on 24 September 2009. Retrieved 24 September 2009.
188. (14 July 2014) Adoption of Genetically Engineered Crops in the U.S.USDA, Economic Research Service, Retrieved 6 August 2014
189. James, Clive (2007). "Executive Summary". G lobal Status of Commercialized Biotech/GM Crops: 2007. ISAAA Briefs. 37. The International Service for the Acquisition of Agri-biotech Applications (ISAAA). ISBN 978-1-892456-42-7. OCLC 262649526. Archived from the original on 6 June 2008. Retrieved 24

September 2009.

190. "Roundup Ready soybean trait patent nears expiration in 2014". Hpj.com. Retrieved 2016-06-06.

191. Acreage NASS National Agricultural Statistics Board annual report, 30 June 2010. Retrieved 23 July 2010.

192. USA:Cultivation of GM Plants in 2009, Maize, soybean, cotton: 88 percent genetically modified GMO Compass. Retrieved 25 July 2010.

193. Fernandez-Cornejo, Jorge (5 July 2012) Adoption of Genetically Engineered Crops in the U.S. – Recent Trends USDA Economic Research Service. Retrieved 29 September 2012

194. Linda Bren. "Genetic Engineering: The Future of Foods?".

195. Lemaux, Peggy (19 February 2008). "Genetically Engineered Plants and Foods: A Scientist's Analysis of the Issues (Part I)". Annual Review of Plant Biology. 59: 771–812. doi:10.1146/annurev.arplant.58.032806.103840. PMID 18284373. Retrieved 9 May 2009.

196. Spain, Bt maize prevails GMO Compass, 31 March 2010. Retrieved 10 August 2010.

197. GM plants in the EU in 2009 Field area for Bt maize decreasesGMO Compass, 29 March 2010. Retrieved 10 August 2010.

198. "EU GMO ban was illegal, WTO rules". Euractiv.com. 12 May 2006. Retrieved 5 January 2010.

199. "GMO Update: US-EU Biotech Dispute; EU Regulations; Thailand". International Centre for Trade and Sustainable Development. Retrieved 5 January 2010.

200. "Genetically Modified Food and Feed". Retrieved 5 January 2010.

201. http://sustainablepulse.com/2015/10/22/gm-crops-now-banned-in-36-countries-worldwide-sustainable-pulse-research/#.Vqkjw098Nbp

202. http://sustainablepulse.com/2015/10/04/gm-crop-bans-confirmed-in-19-eu-countries/#.Vqkkjk98Nbp

203. Paull, John (2015) The threat of genetically modified organisms (GMOs) to organic agriculture: A case study update, Agriculture & Food, 3: 56-63.

204. Nicolia, Alessandro; Manzo, Alberto; Veronesi, Fabio; Rosellini, Daniele (2013). "An overview of the last 10 years of genetically engineered crop safety research" (PDF). Critical Reviews in Biotechnology. 34 (1): 1–12. doi:10.3109/07388551.2013.823595. PMID 24041244.

205. "State of Food and Agriculture 2003–2004. Agricultural Biotechnology: Meeting the Needs of the Poor. Health and environmental impacts of transgenic crops".

206. Domingo, José L.; Bordonaba, Jordi Giné (2011). "A literature review on the safety assessment of genetically modified plants" (PDF). Environment International. 37 (4): 734–742. doi:10.1016/j.envint.2011.01.003. PMID 21296423. "Statement by the AAAS Board of Directors On Labeling of Genetically Modified Foods" (PDF). Pinholster, Ginger (October 25, 2012). "AAAS Board of Directors: Legally

Mandating GM Food Labels Could "Mislead and Falsely Alarm Consumers"". American Association for the Advancement of Science. Retrieved February 8, 2016.

207. "A decade of EU-funded GMO research (2001–2010)" (PDF). Directorate-General for Research and Innovation. Biotechnologies, Agriculture, Food. European Commission, European Union. 2010. doi:10.2777/97784. ISBN 978-92-79-16344-9. Retrieved February 8, 2016.

208. "AMA Report on Genetically Modified Crops and Foods (online summary)". Some medical organizations, including the British Medical Association, advocate further caution based upon the precautionary principle:

209. Scott, Sydney E.; Inbar, Yoel; Rozin, Paul (2016). "Evidence for Absolute Moral Opposition to Genetically Modified Food in the United States" (PDF). Perspectives on Psychological Science. 11 (3): 315–324. doi:10.1177/1745691615621275. PMID 27217243.

210. American Medical Association (2012). Report 2 of the Council on Science and Public Health: Labeling of Bioengineered Foods

211. United States Institute of Medicine and National Research Council (2004). Safety of Genetically Engineered Foods: Approaches to Assessing Unintended Health Effects. National Academies Press. Free full-text. National Academies Press. See pp11ff on need for better standards and tools to evaluate GM food.

212. Key S, Ma JK, Drake PM; Ma; Drake (June 2008). "Genetically modified plants and human health". J R Soc Med. 101 (6): 290–8. doi:10.1258/jrsm.2008.070372. PMC 2408621 . PMID 18515776.

213. Andrew Pollack for the New York Times. "An Entrepreneur Bankrolls a Genetically Engineered Salmon" Published: May 21, 2012. Accessed September 3, 2012

214. Domingo, José L.; Bordonaba, Jordi Giné (2011). "A literature review on the safety assessment of genetically modified plants" (PDF). Environment International. 37 (4): 734–742. doi:10.1016/j.envint.2011.01.003. PMID 21296423.

215. Krimsky, Sheldon (2015). "An Illusory Consensus behind GMO Health Assessment" (PDF). Science, Technology, & Human Values. 40 (6): 1–32. doi:10.1177/0162243915598381.

216. Panchin, Alexander Y.; Tuzhikov, Alexander I. (2016). "Published GMO studies find no evidence of harm when corrected for multiple comparisons". Critical Reviews in Biotechnology. 37 (2): 1–5. doi:10.3109/07388551.2015.1130684. PMID 26767435.

CHAPTER 8

Chloroplast Capture

Chapter Contents

Plants use energy from the sun in tiny energy factories called chloroplasts. Using chlorophyll in the process called photosynthesis, they convert the sun's energy into storable form in ordered sugar molecules such as glucose. In this way, carbon dioxide from the air and water from the soil in a more disordered state are combined to form the more ordered sugar molecules.

INTRODUCTION

Chloroplasts are organelles, specialized subunits, in plant and algal cells. Their discovery inside plant cells is usually credited to Julius von Sachs (1832–1897), an influential botanist and author of standard botanical textbooks – sometimes called "The Father of Plant Physiology". The main role of chloroplasts is to conduct photosynthesis, where the photosynthetic pigment chlorophyll captures the energy from sunlight and converts it and stores it in the energy-storage molecules ATP and NADPH while freeing oxygen from water. They then use the ATP and NADPH to make organic molecules from carbon dioxide in a process known as the Calvin cycle. Chloroplasts carry out a number of other functions, including fatty acid synthesis, much amino acid synthesis, and the immune response in plants. The number of chloroplasts per cell varies from one, in unicellular algae, up to 100 in plants like Arabidopsis and wheat.

A chloroplast is a type of organelle known as a plastid, characterized by its high concentration of chlorophyll. Other plastid types, such as the leucoplast and the chromoplast, contain little chlorophyll and do not carry out photosynthesis.

Chloroplasts are highly dynamic—they circulate and are moved around within plant cells, and occasionally pinch in two to reproduce. Their behavior is strongly influenced by environmental factors like light color and intensity. Chloroplasts, like mitochondria, contain their own DNA, which is thought to be inherited from their ancestor—a photosynthetic cyanobacterium that was engulfed by an early eukaryotic cell. Chloroplasts cannot be made by the plant cell and must be inherited by each daughter cell during cell division.

KNOWN CASES OF CHLOROPLAST CAPTURE

Gymnosperm

- *Juniperus* (Cupressaceae)
- *Pinus* (Pinaceae)

Juniper

Junipers are coniferous plants in the genus Juniperus of the cypress family Cupressaceae. Depending on taxonomic viewpoint, between 50 and 67 species of juniper are widely distributed throughout the Northern Hemisphere, from the Arctic, south to tropical Africa, from Ziarat, Pakistan east to eastern Tibet in the Old World, and in the mountains of Central America. The highest-known Juniper forest occurs at an altitude of 16,000 feet (4,900 m) in south-eastern Tibet and the northern Himalayas, creating one of the highest tree-lines on earth.

Description

Cones and leaves of *Juniperus communis*

Junipers vary in size and shape from tall trees, 20–40 m (66–131 ft) tall, to columnar or low spreading shrubs with long trailing branches. They are evergreen with needle-like and/or scale-like leaves. They can be either monoecious or dioecious. The female seed cones are very distinctive, with fleshy, fruit-like coalescing scales which fuse together to form a "berry"-like structure, 4–27 mm (0.16–1.06 in) long, with 1–12 unwinged, hard-shelled seeds. In some species these "berries" are red-brown or orange but in most they are blue; they are often aromatic and can be used as a spice. The seed maturation time varies between species from 6–18 months after pollination. The male cones are similar to those of other Cupressaceae, with 6–20 scales.

In zones 7 through 10, junipers can bloom and release pollen several times each year. A few species of juniper bloom in autumn, while most species pollinate from early winter until late spring.

Detail of *Juniperus chinensis* shoots, with juvenile (needle-like) leaves (left), and adult scale leaves and immature male cones (right)

Many junipers (e.g. *J. chinensis*, *J. virginiana*) have two types of leaves: seedlings and some twigs of older trees have needle-like leaves 5–25 mm (0.20–0.98 in) long; and the leaves on mature plants are (mostly) tiny (2–4 mm (0.079–0.157 in)), overlapping and scale-like. When juvenile foliage occurs on mature plants, it is most often found on shaded shoots, with adult foliage in full sunlight. Leaves on fast-growing 'whip' shoots are often intermediate between juvenile and adult.

In some species (e. g. *J. communis*, *J. squamata*), all the foliage is of the juvenile needle-like type, with no scale leaves. In some of these (e.g. *J. communis*), the needles are jointed at the base, in others (e.g. *J. squamata*), the needles merge smoothly with the stem, not jointed.

The needle-leaves of junipers are hard and sharp, making the juvenile foliage very prickly to handle. This can be a valuable identification feature in seedlings, as the otherwise very similar juvenile foliage of cypresses (*Cupressus, Chamaecyparis*) and other related genera is soft and not prickly.

Juniper is the exclusive food plant of the larvae of some Lepidoptera species including *Bucculatrix inusitata* and juniper carpet, and is also eaten by the larvae of other Lepidoptera species such as *Chionodes electella*, *Chionodes viduella*, juniper pug and pine beauty; those of the tortrix moth *C. duplicana* feed on the bark around injuries or canker.

CLASSIFICATION

Juniper needles, magnified. Left, *Juniperus communis* (*Juniperus* sect. *Juniperus*; note needles 'jointed' at base). Right, *Juniperus chinensis* (*Juniperus* sect. *Sabina*; note needles merging smoothly with the stem, not jointed at base).

Juniperus phoenicea on El Hierro, Canary Islands

Juniperus virginiana in October laden with ripe cones.

Cones and seeds.

Juniperus occidentalis var. *australis*, eastern Sierra Nevada, Rock Creek Canyon, California.

Juniperus communis wood pieces, with a U.S. penny for scale, showing the narrow growth rings of the species.

The number of juniper species is in dispute, with two recent studies giving very different totals, Farjon (2001) accepting 52 species, and Adams (2004) accepting 67 species. The junipers are divided into several sections, though (particularly among the scale-leaved species) which species belong to which sections is still far from clear, with research still on-going. The section *Juniperus* is an obvious monophyletic group though.

- ***Juniperus* sect. *Juniperus*:** Needle-leaf junipers. The adult leaves are needle-like, in whorls of three, and jointed at the base (see below right).
- *Juniperus* sect. *Juniperus* subsect. *Juniperus*: Cones with 3 separate seeds; needles with one stomatal band.
- *Juniperus communis* - Common Juniper
- *Juniperus communis* subsp. *alpina* - Alpine Juniper
- *Juniperus conferta* - Shore Juniper (syn. *J. rigida* var. *conferta*)
- *Juniperus rigida* - Temple Juniper or Needle Juniper
- *Juniperus* sect. *Juniperus* subsect. *Oxycedrus*: Cones with 3 separate seeds; needles with two stomatal bands.
- *Juniperus brevifolia* - Azores Juniper
- *Juniperus cedrus* - Canary Islands Juniper
- *Juniperus deltoides* - Eastern Prickly Juniper
- *Juniperus formosana* - Chinese Prickly Juniper
- *Juniperus lutchuensis* - Ryukyu Juniper
- *Juniperus navicularis* - Portuguese Prickly Juniper
- *Juniperus oxycedrus* - Western Prickly Juniper or Cade Juniper
- *Juniperus macrocarpa* (*J. oxycedrus* subsp. *macrocarpa*) - Large-berry Juniper
- *Juniperus* sect. *Juniperus* subsect. *Caryocedrus*: Cones with 3 seeds fused together; needles with two stomatal bands.

- *Juniperus drupacea* - Syrian Juniper
- ***Juniperus* sect. *Sabina*:** Scale-leaf junipers. The adult leaves are mostly scale-like, similar to those of *Cupressus* species, in opposite pairs or whorls of three, and the juvenile needle-like leaves are not jointed at the base (including in the few that have only needle-like leaves; see below right). Provisionally, all the other junipers are included here, though they form a paraphyletic group.
- Old World species
- *Juniperus chinensis* - Chinese Juniper
- *Juniperus chinensis* var. *sargentii* - Sargent's Juniper (Japanese: 深山柏槇)
- *Juniperus chinensis* L. var. *tsukusiensis Masummune* (Chinese: 清水圓柏)
- *Juniperus chinensis Kaizuka* (Japanese: 貝塚伊吹)
- *Juniperus chinensis* var. *Procumbens* (Japanese: 這柏槇)
- *Juniperus chinensis Globosa* (Japanese: 玉伊吹)
- *Juniperus chinensis Aurea'* (Japanese: 金伊吹)
- *Juniperus convallium* - Mekong Juniper
- *Juniperus excelsa* - Greek Juniper
- *Juniperus excelsa polycarpos* - Persian Juniper
- *Juniperus foetidissima* - Stinking Juniper
- *Juniperus indica* - Black Juniper
- *Juniperus komarovii* - Komarov's Juniper
- *Juniperus phoenicea* - Phoenicean Juniper
- *Juniperus pingii* - Ping Juniper
- *Juniperus pingii* var. *chengii*
- *Juniperus pingii* var. *miehei*
- Juniperus pingii var. *wilsonii*
- *Juniperus procera* - East African Juniper
- *Juniperus procumbens* - Ibuki Juniper
- *Juniperus pseudosabina* - Xinjiang Juniper
- *Juniperus recurva* - Himalayan Juniper
- *Juniperus recurva* var. *butanica*
- *Juniperus recurva* var. *coxii* - Cox's Juniper
- *Juniperus sabina* - Savin Juniper
- *Juniperus sabina* var. *davurica* - Daurian Juniper
- *Juniperus saltuaria* - Sichuan Juniper
- *Juniperus semiglobosa* - Russian Juniper
- *Juniperus squamata* - Flaky Juniper
- *Juniperus thurifera* - Spanish Juniper

- *Juniperus tibetica* - Tibetan Juniper
- *Juniperus wallichiana* - Himalayan Black Juniper
- New World species
- *Juniperus angosturana* - Mexican One-seed Juniper
- *Juniperus ashei* - Ashe Juniper
- *Juniperus arizonica* - Synonym: Juniperus coahuilensis var. arizonica or Juniperus erythrocarpa var. coahuilensis. common names: Redberry Juniper, Roseberry Juniper
- *Juniperus barbadensis* - West Indies Juniper
- *Juniperus bermudiana* - Bermuda Juniper
- *Juniperus blancoi* - Blanco's Juniper
- *Juniperus californica* - California Juniper
- *Juniperus coahuilensis* - Coahuila Juniper
- *Juniperus comitana* - Comitán Juniper
- *Juniperus deppeana* - Alligator Juniper
- *Juniperus durangensis* - Durango Juniper
- *Juniperus flaccida* - Mexican Weeping Juniper
- *Juniperus gamboana* - Gamboa Juniper
- *Juniperus horizontalis* - Creeping Juniper
- *Juniperus jaliscana* - Jalisco Juniper
- *Juniperus maritima* - Seaside Juniper
- *Juniperus monosperma* - One-seed Juniper
- *Juniperus monticola* - Mountain Juniper
- *Juniperus occidentalis* - Western Juniper
- *Juniperus occidentalis* subsp. *australis* - Sierra Juniper
- *Juniperus osteosperma* - Utah Juniper
- *Juniperus pinchotii* - Pinchot Juniper
- *Juniperus saltillensis* - Saltillo Juniper
- *Juniperus scopulorum* - Rocky Mountain Juniper
- *Juniperus standleyi* - Standley's Juniper
- *Juniperus virginiana* - Eastern Juniper (Eastern Redcedar)
- *Juniperus virginiana* subsp. *silicicola* - Southern Juniper
- *Juniperus zanonii* (proposed)

Cultivation and Uses

Juniper (*Juniperus osteosperma* and *scopulorum*) essential oil

Juniper berries are a spice used in a wide variety of culinary dishes and best known for the primary flavoring in gin (and responsible for gin's name, which is a shortening of the Dutch word for juniper, *genever*). Juniper berries are also used as the primary flavor in the liquor Jenever and sahti-style of beers. Juniper berry sauce is often a popular flavoring choice for quail, pheasant, veal, rabbit, venison and other meat dishes.

Many of the earliest prehistoric people lived in or near juniper forests which furnished them food, fuel, and wood for shelter or utensils. Many species, such as *J. chinensis* (Chinese Juniper) from eastern Asia, are extensively used in landscaping and horticulture, and as one of the most popular species for use in bonsai. It is also a symbol of longevity, strength, athleticism, and fertility.

Some junipers are susceptible to *Gymnosporangium* rust disease, and can be a serious problem for those people growing apple trees, the alternate host of the disease.

Some junipers are given the common name "cedar," including *Juniperus virginiana*, the "red cedar" that is used widely in cedar drawers. "Eastern redcedar" is the correct name for *J. virginiana*. The lack of space between the words "red" and "cedar" indicate that this species is not a true cedar, Cedrus.

In Morocco, the tar (gitran) of the arar tree (Juniperus phoenicea) is applied in dotted patterns on bisque drinking cups. Gitran makes the water more fragrant and is said to be good for the teeth.

Some Indigenous peoples, such as the Dineh, have traditionally used juniper to treat diabetes. Animal studies have shown that treatment with juniper may retard the development of streptozotocin-induced diabetes in mice. Native Americans have also used juniper berries as a female contraceptive. The 17th Century herbalist physician Nicholas Culpeper recommended the ripened berries for conditions such as asthma and sciatica, as well as to speed childbirth.

Juniper is one of the plants used in Scottish and Gaelic Polytheist saining rites, such as those performed at Hogmanay (New Year), where the smoke of burning juniper is used to cleanse, bless and protect the household and its inhabitants.

Juniper berries are steam distilled to produce an essential oil that may vary from colorless to yellow or pale green. Some of its chemical components are alpha pinene, cadinene, camphene and terpineol. Leaves and twigs of Juniperus virginiana are steam distilled to produce oil of juniper. Middle Tennessee and adjacent northern Alabama and southern Kentucky are the centers for this activity. The U.S. Forest Service has provided plans for the apparatus required. This work is typically done during periods of cold weather to reduce the loss of essential oil to evaporation, which is greater in warmer weather, and to take advantage of a time of year when labor might be more readily available.

Juniper in weave is a traditional cladding technique used in Northern Europe, e.g. at Havrå, Norway.

Allergenic Potential

In drier areas, juniper pollen easily becomes airborne and can be inhaled into the lungs. This pollen can also irritate the skin and cause contact dermatitis. Cross-allergenic reactions are common between juniper pollen and the pollen of all species of cypress.

Monoecious juniper plants are highly allergenic, with an OPALS allergy scale rating of 9 out of 10. Completely male juniper plants have an OPALS rating of 10, and release abundant amounts of pollen. Conversely, all-female juniper plants have an OPALS rating of 1, and are considered "allergy-fighting".

Pine

A pine is any conifer in the genus Pinus, of the family Pinaceae. Pinus is the sole genus in the subfamily Pinoideae. The Plant List compiled by the Royal Botanic Gardens, Kew and Missouri Botanical Garden accepts 126 species names of pines as current, together with 35 unresolved species and many more synonyms.

Etymology

The modern English name *pine* derives from Latin *pinus*, which some have traced to the Indo-European base *pīt- 'resin' (source of English pituitary). Before the 19th century, pines were often referred to as firs (from Old Norse *fura*, by way of Middle English*firre*). In some European languages, Germanic cognates of the Old Norse name are still in use for pines—in Danish *fyr*, in Norwegian*fura/fure/furu*, Swedish *fura/furu*, Dutch *vuren*, and German *Föhre*—but in modern English, *fir* is now restricted to fir (*Abies*) and Douglas fir (*Pseudotsuga*).

Pine Forest in Vagamon, Southern Western Ghats, Kerala (India)

Taxonomy, Nomenclature and Codification

Pines are gymnosperms. The genus is divided into three subgenera, which can be distinguished by cone, seed, and leaf characters:

- *Pinus* subg. *Pinus*, the yellow, or hard pine group, generally with harder wood and two or three needles per fascicle
- *Pinus* subg. *Ducampopinus*, the foxtail or pinyon group
- *Pinus* subg. *Strobus*, the white, or soft pine group, generally with softer wood and five needles per fascicle

Distribution

A Khasi pine in Benguet, Philippines

Huangshan pine (*Pinus hwangshanensis*), Anhui, China

Most regions of the Northern Hemisphere (see List of pines by region) host some native species of pines. One species (Sumatran pine) crosses the equator in Sumatra to 2°S. In North America, various species occur in regions at latitudes from as far north as 66°N to as far south as 12°N.

Various species have been introduced to temperate and subtropical regions of both hemispheres, where they are grown as timber or cultivated as ornamental plants in parks and gardens. A number of such introduced species have become invasive and threaten native ecosystems.

DESCRIPTION

Ancient *Pinus longaeva*, Nevada, USA

Pine trees are evergreen, coniferous resinous trees (or, rarely, shrubs) growing 3–80 m (10–260 ft) tall, with the majority of species reaching 15–45 m (50–150 ft) tall. The smallest are Siberian dwarf pine and Potosi pinyon, and the tallest is a 81.79 m (268.35 ft) tall ponderosa pine located in southern Oregon's Rogue River-Siskiyou National Forest.

Bark chips

The bark of most pines is thick and scaly, but some species have thin, flaky bark. The branches are produced in regular "pseudo whorls", actually a very tight spiral but appearing like a ring of branches arising from the same point. Many pines are *uninodal*, producing just one such whorl of branches each year, from buds at the tip of the year's new shoot, but others are *multinodal*, producing two or more whorls of branches per year. The spiral growth of branches, needles, and cone scales are arranged in Fibonacci number ratios. The new spring shoots are sometimes called "candles"; they are covered in brown or whitish bud scales and point upward at first, then later turn green and spread outward. These "candles" offer foresters a means to evaluate fertility of the soil and vigour of the trees.

Pines are long-lived, and typically reach ages of 100–1,000 years, some even more. The longest-lived is the Great Basin bristlecone pine, *Pinus longaeva*. One individual of this species, dubbed "Methuselah", is one of the world's oldest living organisms at around 4,600 years old. This tree can be found in the White Mountains of California. An older tree, now cut down, was dated at 4,900 years old. It was discovered in a grove beneath Wheeler Peak and it is now known as "Prometheus" after the Greek immortal.

Foliage

Illustration of needles, cones, and seeds of Scots pine (*Pinus sylvestris*)

Pines have four types of leaf:

- *Seed leaves* (cotyledons) on seedlings, born in a whorl of 4–24.
- *Juvenile leaves*, which follow immediately on seedlings and young plants, 2–6 cm long, single, green or often blue-green, and arranged spirally on the shoot. These are produced for six months to five years, rarely longer.
- *Scale leaves*, similar to bud scales, small, brown and non-photosynthetic, and arranged spirally like the juvenile leaves.
- *Needles*, the adult leaves, which are green (photosynthetic), bundled in clusters (*fascicles*) of 1–6, commonly 2–5, needles together, each fascicle produced

from a small bud on a dwarf shoot in the axil of a scale leaf. These bud scales often remain on the fascicle as a basal sheath. The needles persist for 1.5–40 years, depending on species. If a shoot is damaged (e.g. eaten by an animal), the needle fascicles just below the damage will generate a bud which can then replace the lost leaves.

Cones

Pines are mostly monoecious, having the male and female cones on the same tree, though a few species are sub-dioecious, with individuals predominantly, but not wholly, single-sex. The male cones are small, typically 1–5 cm long, and only present for a short period (usually in spring, though autumn in a few pines), falling as soon as they have shed their pollen. The female cones take 1.5–3 years (depending on species) to mature after pollination, with actual fertilization delayed one year. At maturity the female cones are 3–60 cm long. Each cone has numerous spirally arranged scales, with two seeds on each fertile scale; the scales at the base and tip of the cone are small and sterile, without seeds. The seeds are mostly small and winged, and are anemophilous (wind-dispersed), but some are larger and have only a vestigial wing, and are bird-dispersed (see below). At maturity, the cones usually open to release the seeds, but in some of the bird-dispersed species (e.g. whitebark pine), the seeds are only released by the bird breaking the cones open. In others, the seeds are stored in closed ("serotinous") cones for many years until an environmental cue triggers the cones to open, releasing the seeds. The most common form of serotiny is pyriscence, in which a resin binds the cones shut until melted by a forest fire.

Ecology

A prescribed fire in a European black pine (*Pinus nigra*) woodland, Portugal

Pines grow well in acid soils, some also on calcareous soils; most require good soil drainage, preferring sandy soils, but a few (e.g. lodgepole pine) will tolerate poorly drained wet soils. A few are able to sprout after forest fires (e.g. Canary Island pine). Some species of pines (e.g. bishop pine) need fire to regenerate, and their populations slowly decline under fire suppression regimes. Several species are adapted to extreme conditions imposed by elevation and latitude (e.g. Siberian dwarf pine, mountain pine, whitebark pine and the bristlecone pines). The pinyon pines and a number of others, notably Turkish pine and gray pine, are particularly well adapted to growth in hot, dry semi-desert climates.

The seeds are commonly eaten by birds, such as grouse, crossbills, jays, nuthatches, siskins, woodpeckers, and by squirrels. Some birds, notably the spotted nutcracker, Clark's nutcracker and Pinyon jay, are of importance in distributing pine seeds to new areas. Pine needles are sometimes eaten by some Lepidoptera (butterfly and moth) species (see list of Lepidoptera that feed on pines), the Symphytan species pine sawfly, and goats.

Pine pollen may play an important role in the functioning of detrital food webs. Nutrients from pollen aid detritivores in development, growth and maturation, and may enable fungi to decompose nutritionally scarce litter. Pine pollen is also involved in moving plant matter between terrestrial and aquatic ecosystems.

Use

Logging *Pinus ponderosa*, Arizona, USA

Pinus sylvestris prepared for transport, Hungary

Pine cone

Tongue and groove solid pine flooring

Pines are among the most commercially important tree species valued for their timber and wood pulp throughout the world. In temperate and tropical regions, they are fast-growing softwoods that will grow in relatively dense stands, their acidic decaying needles inhibiting the sprouting of competing hardwoods. Commercial pines are grown in plantations for timber that is denser, more resinous, and therefore more durable than spruce (*Picea*). Pine wood is widely used in high-value carpentry items such as furniture, window frames, panelling, floors and roofing, and the resin of some species is an important source of turpentine.

Many pine species make attractive ornamental plantings for parks and larger gardens with a variety of dwarf cultivars being suitable for smaller spaces. Pines are also commercially grown and harvested for Christmas trees. Pine cones, the largest and most durable of all conifer cones, are craft favorites. Pine boughs, appreciated especially in wintertime for their pleasant smell and greenery, are popularly cut for decorations. A number of species are attacked by nematodes, causing pine wilt disease, which can kill some quickly. Pine needles are also used for making decorative articles like baskets, trays, pots, etc. This Native American skill is now being replicated across the world. Pine needle handicrafts are made in the US, Canada, Mexico, Nicaragua and India. Pine needles serve as food for various Lepidoptera. See List of Lepidoptera that feed on pines.

Because pines have no insect or decay resistant qualities after logging, they are generally recommended for construction purposes as indoor use only (ex. indoor drywall framing). This wood left outside can be expected to last no more than 12–18 months depending on the local climate. It is commonly referred to by several different names which include North American timber, SPF (spruce, pine, fir) and whitewood.

Farming

When grown for sawing timber, pine plantations can be harvested after 30 years, with some stands being allowed to grow up to 50 (as the wood value increases more quickly as the trees age). Imperfect trees (such as those with bent trunks or forks, smaller trees, or diseased trees) are removed in a “thinning” operation every five to ten years. Thinning allows the best trees to grow much faster, because it prevents weaker trees from competing for sunlight, water, and nutrients. Young trees removed during thinning are used for pulpwood, while most older ones are good enough for saw timber.

The final wood quality can be improved by pruning small branches at ages 5, 7, and 9. Pruning usually goes up to a height of 6 metres (20 ft). This results in smooth timber with no knots, which is considerably more valuable.

A 30-year-old commercial pine tree grown in good conditions will be about 0.3 metres (1.0 ft) in diameter and about 20 metres (66 ft) high. After 50 years, the same tree will be about 0.5 metres (1.6 ft) in diameter and 25 metres (82 ft) high, and its wood will be worth about 7 times as much as the 30-year-old tree.

Trees are planted 3 to 4 metres apart, or about 1000 per hectare (100,000 per square kilometer).

Food

Some species have large seeds, called pine nuts, that are harvested and sold for cooking and baking. They are an essential ingredient of *Pesto alla genovese*.

The soft, moist, white inner bark (cambium) found clinging to the woody outer bark is edible and very high in vitamins A and C. It can be eaten raw in slices as a snack or dried and ground up into a powder for use as an ersatz flour or thickener in stews, soups, and other foods, such as bark bread. Adirondack Indians got their name from the Mohawk Indian word *atirú:taks*, meaning "tree eaters".

A tea made by steeping young, green pine needles in boiling water (known as "tallstrunt" in Sweden) is high in vitamins A and C.

Edible seeds of the Korean pine (*Pinus koraiensis*)

Folk Medicine

Pine has been listed as one of the 38 substances used to prepare Bach flower remedies, a kind of alternative medicine promoted for its effect on health. However, according to Cancer Research UK, "there is no scientific evidence to prove that flower remedies can control, cure or prevent any type of disease, including cancer".

Angiosperm

- *Quercus* (Fagaceae)
- *Ruppia* (Ruppiaceae)
- *Heuchera* (Saxifragaceae)

OAK

An **oak** is a tree or shrub in the genus ***Quercus*** of the beech family, Fagaceae. There are approximately 600 extant species of oaks. The common name "oak" also appears in the names of species in related genera, notably *Lithocarpus* (stone oaks), as well as in those of unrelated species such as *Grevillea robusta* (silky oaks) and the Casuarinaceae (she-oaks). The genus *Quercus* is native to the Northern Hemisphere, and includes deciduous and evergreen species extending from cool temperate to tropical latitudes in the Americas, Asia, Europe, and North Africa. North America contains the largest number of oak species, with approximately 90 occurring in the United States, while Mexico has 160 species of which 109 are endemic. The second greatest center of oak diversity is China, which contains approximately 100 species.

Oaks have spirally arranged leaves, with lobate margins in many species; some have serrated leaves or entire leaf with smooth margins. Also, the acorns contain tannic acid, as do the leaves, which helps to guard from fungi and insects. Many deciduous species are marcescent, not dropping dead leaves until spring. In spring, a single oak tree produces both male flowers (in the form of catkins) and small female flowers. The fruit is a nut called an acorn, borne in a cup-like structure known as a cupule; each acorn contains one seed (rarely two or three) and takes 6–18 months to mature, depending on their species. The live oaks are distinguished for being evergreen, but are not actually a distinct group and instead are dispersed across the genus.

Classification

The oak tree is a flowering plant. Oaks may be divided into two genera (sometimes referred to as subgenera) and a number of sections:

Genus *Quercus*

Oak at Schönderling

The genus *Quercus* is divided into the following sections:

- Sect. *Quercus* (synonyms *Lepidobalanus* and *Leucobalanus*), the white oaks of Europe, Asia and North America. Styles are short; acorns mature in 6 months and taste sweet or slightly bitter; the inside of an acorn shell is hairless. The leaves mostly lack a bristle on their lobe tips, which are usually rounded. The type species is Quercus robur.
- Sect. *Mesobalanus*, Hungarian oak and its relatives of Europe and Asia. Styles long; acorns mature in about 6 months and taste bitter; the inside of this acorn's shell is hairless. The section *Mesobalanus* is closely related to section *Quercus* and sometimes included in it.
- Sect. *Cerris*, the Turkey oak and its relatives of Europe and Asia. Styles long; acorn mature in 18 months and taste very bitter. The inside of the acorn's shell is hairless. Its leaves typically have sharp lobe tips, with bristles at the lobe tip.
- Sect. *Protobalanus*, the canyon live oak and its relatives, in southwest United States and northwest Mexico. Styles short, acorns mature in 18 months and taste very bitter. The inside of the acorn shell appears woolly. Leaves typically have sharp lobe tips, with bristles at the lobe tip.
- Sect. *Lobatae* (synonym *Erythrobalanus*), the red oaks of North America, Central America and northern South America. Styles long; acorns mature in 18 months and taste very bitter. The inside of the acorn shell appears woolly. The actual nut is encased in a thin, clinging, papery skin. Leaves typically have sharp lobe tips, with spiny bristles at the lobe.

Genus *Cyclobalanopsys*

Old oak tree on the shore of Lake Koluvere, Estonia.

- The ring-cupped oaks of eastern and southeastern Asia. Evergreen trees growing 10–40 m (33–131 ft) tall. They are distinct from subgenus *Quercus* in that they have acorns with distinctive cups bearing concrescent rings of scales; they commonly also have densely clustered acorns, though this does not apply to all of the species. IUCN, ITIS, Encyclopedia of Life and *Flora of China* treats *Cyclobalanopsis* as a distinct genus, but some taxonomists consider it a subgenus of *Quercus*. It contains about 150 species. Species of *Cyclobalanopsis* are common in the evergreen subtropical laurel forests which extend from southern Japan, southern Korea, and Taiwan across southern China and northern Indochina to the eastern Himalayas, in association with trees of genus *Castanopsis* and the laurel family (Lauraceae).

Hybridization

A hybrid white oak, possibly *Quercus stellata* × *Q. muhlenbergii*

Interspecific hybridization is quite common among oaks but usually between species within the same section only and most common in the white oak group (subgenus *Quercus*, section *Quercus*; see **List of *Quercus* species**). Inter-section hybrids, except between species of sections *Quercus* and *Mesobalanus*, are unknown. Recent systematic studies appear to confirm a high tendency of *Quercus* species to hybridize because of a combination of factors. White oaks are unable to discriminate against pollination by other species in the same section. Because they are wind pollinated and they have weak internal barriers to hybridization, hybridization produces functional seeds and fertile hybrid offspring. Ecological stresses, especially near habitat margins, can also cause a breakdown of mate recognition as well as a reduction of male function (pollen quantity and quality) in one parent species.

Frequent hybridization among oaks has consequences for oak populations around the world; most notably, hybridization has produced large populations of hybrids with copious amounts of introgression, and the evolution of new species. Frequent hybridization and high levels of introgression have caused different species in the same populations to share up to 50% of their genetic information. Having high rates of hybridization and introgression produces genetic data that often does not differentiate between two clearly morphologically distinct species, but instead differentiates populations. Numerous hypotheses have been proposed to explain how oak species are able to remain morphologically and ecologically distinct with such high levels of gene flow, but the phenomenon is still largely a mystery to botanists.

The Fagaceae, or beech family, to which the oaks belong, is a very slow evolving clade compared to other angiosperms, and the patterns of hybridization and introgression in *Quercus* pose a great challenge to the concept of a species since a species is often defined as a group of "actually or potentially interbreeding populations which are reproductively isolated from other such groups." By this definition, many species of *Quercus* would be lumped together according to their geographic and ecological habitat, despite clear distinctions in morphology and, to a large extent, genetic data.

Uses

Heart of oak beams of the frame of Saint-Girons church in Monein, France

Oak wood has a density of about 0.75 g/cm^3 (0.43 oz/cu in) creating great strength and hardness. The wood is very resistant to insect and fungal attack because of its high tannin content. It also has very appealing grain markings, particularly when quartersawn. Oak planking was common on high status Viking longships in the 9th and 10th centuries. The wood was hewn from green logs, by axe and wedge, to produce radial planks, similar to quarter-sawn timber. Wide, quarter-sawn boards of oak have been prized since the Middle Ages for use in interior panelling of prestigious buildings such as the debating chamber of the House of Commons in London and in the construction of fine furniture. Oak wood, from *Quercus robur* and *Quercus petraea*, was used in Europe for the construction of ships, especially naval men of war, until the 19th century, and was the principal timber used in the construction of European timber-framed buildings. Today oak wood is still commonly used for furniture making and flooring, timber frame buildings, and for veneer production. Barrels in which wines, sherry, and spirits such as brandy, Irish whiskey, Scotch whisky and Bourbon whiskey are aged are made from European and American oak. The use of oak in wine can add many different dimensions to wine based on the type and style of the oak. Oak barrels, which may be charred before use, contribute to the colour, taste, and aroma of the contents, imparting a desirable oaky vanillin flavour to these drinks. The great dilemma for wine producers is to choose between French and American oakwoods. French oaks (*Quercus robur*, *Q. petraea*) give the wine greater refinement and are chosen for the best wines since they increase the price compared to those aged in American oak wood. American oak contributes greater texture and resistance to ageing, but produces more powerful wine bouquets. Oak wood chips are used for smoking fish, meat, cheeses, and other foods.

Sherry maturing in oak barrels

Japanese oak is used in the making of professional drums from the manufacturer Yamaha Drums. The higher density of oak gives the drum a brighter and louder tone compared to traditional drum materials such as maple and birch. In hill states of India, besides fuelwood and timber, the local people use oak wood for making agricultural implements. The leaves are used as fodder during lean period and bedding for livestock.

A cross section of the trunk of a cork oak, *Quercus suber*

The bark of the cork oak is used to produce wine stoppers (corks). This species grows in the Mediterranean Sea region, with Portugal, Spain, Algeria, and Morocco producing most of the world's supply.

Of the North American oaks, the northern red oak is the one of the most prized of the red oak group for lumber, much of which is marketed as red oak regardless of the species of origin. It is not good for outdoor use due to its open capillaries unless the wood is treated. If the wood is properly treated with preservatives, it will not rot as quickly as cured white oak heartwood. The closed cell structure of white oaks prevent them from absorbing preservatives. With northern red oak, one can blow air through an end grain piece 10 inches long to make bubbles come out in a glass of water. These openings give fungus easy access when the finish deteriorates. Shumard oak, a member of the red oak subgenus, provides timber which is described as "mechanically superior" to Northern Red oak. Cherrybark oak is another type of red oak which provides excellent timber.

The standard for the lumber of the white oak group – all of which is marketed as white oak – is the Quercus alba. White oak is often used to make wine barrels. The wood of the deciduous pedunculate oak and sessile oak accounts for most of the European oak production, but evergreen species, such as Holm oak and cork oak also produce valuable timber.

The bark of the White Oak is dried and used in medical preparations. Oak bark is also rich in tannin, and is used by tanners for tanning leather. Acorns are used for making flour or roasted for acorn coffee.

Oak galls were used for centuries as a main ingredient in iron gall ink, a kind of manuscript ink, harvested at a specific time of year. In Korea, oak bark is used to make shingles for traditional roof construction.

Oak has been listed as one of the 38 substances used to prepare Bach flower remedies, a kind of alternative medicine promoted for its effect on health. However, according to Cancer Research UK, "there is no scientific evidence to prove that flower remedies can control, cure or prevent any type of disease, including cancer".

Biodiversity and Ecology

Oak forest in Estonia.

Oak on sandy earth.

Oaks are keystone species in a wide range of habitats from Mediterranean semi-desert to subtropical rainforest. For example, oak trees are important components of hardwood forests, and certain species are particularly known to grow in associations with members of the Ericaceae in oak-heath forests. A number of kinds of truffles, including the two well known varieties, the black Périgord truffle and the white Piedmont truffle, have symbiotic relationships with oak trees. The European pied flycatcher is an example of an animal species that often depends upon oak trees.

Many species of oaks are under threat of extinction in the wild, largely due to land use changes, livestock grazing and unsustainable harvesting. For example, over the past 200 years, large areas of oak forest in the highlands of Mexico, Central America and the northern Andes have been cleared for coffee plantations and cattle ranching. There is a continuing threat to these forests from exploitation for timber, fuelwood and charcoal. In the USA, entire oak ecosystems have declined due to a combination of factors still imperfectly known, but thought to include fire suppression, increased consumption of acorns by growing mammal populations, herbivory of seedlings, and introduced pests. In a recent survey, 78 wild oak species have been identified as being in danger of extinction, from a global total of over 500 species.The proportion under threat may be much higher in reality, as there is insufficient information about over 300 species, making it near impossible to form any judgement of their status.

In the Himalayan region of India, oak forests are being invaded by pine forests due to the increase in temperature. The associated species of pine forest may cross frontiers and become new elements of the oak forests.

In eastern North America, rare species of oak trees include scarlet oak (*Quercus coccinea*), chinkapin oak (*Quercus muehlenbergii*), and post oak (*Quercus stellata*).

The mature trees shed varying numbers of acorns annually. Scientist suggest that shedding excess numbers allows the oaks to satiate nut gathering species, improving the chances of germination. One in 10,000 acorns results in an actual tree.

Genetics

Beginning November 1st 2011, a project began to sequence the entire oak genome. The goal of the project is to create a high resolution sequence of the Quercus robur genome, and to study genetic diversity by comparison of the genomes of different species. Current research has compiled genomic data from many different sources and techniques to create a genome map with 89% coverage of the genome. The project is still in the process of annotating this genome.

DISEASES AND PESTS

Oak powdery mildew on pedunculate oak

Sudden oak death (*Phytophthora ramorum*) is a water mould that can kill oaks within just a few weeks. Oak wilt, caused by the fungus *Ceratocystis fagacearum* (a fungus closely related to Dutch elm disease), is also a lethal disease of some oaks, particularly the red oaks (the white oaks can be infected but generally live longer). Other dangers include wood-boring beetles, as well as root rot in older trees which may not be apparent on the outside, often being discovered only when the trees come down in a strong gale. Oak apples are galls on oaks made by the gall wasp. The female kermes scale causes galls to grow on kermes oak. Oaks are used as food plants by the larvae of Lepidoptera (butterfly and moth) species such as the gypsy moth, *Lymantria dispar*, which can defoliate oak and other broadleaved tree species in North America.

A considerable number of galls are found on oak leaves, buds, flowers, roots, etc. Examples are oak artichoke gall, oak marble gall, oak apple gall, knopper gall, and spangle gall.

A number of species of fungus cause powdery mildew on oak species. In Europe the species *Erysiphe alphitoides* is the most common cause.

A new and yet little understood disease of mature oaks, acute oak decline, has been reported in parts of the UK since 2009.

Oak processionary moth (*Thaumetopoea processionea*) has become a serious threat in the UK since 2006. The caterpillars of this species defoliate the trees, and are hazardous to human health; their bodies are covered with poisonous hairs which can cause rashes and respiratory problems.

TOXICITY

The leaves and acorns of the oak tree are poisonous to cattle, horses, sheep, and goats in large amounts due to the toxin tannic acid, and cause kidney damage and gastroenteritis. Symptoms of poisoning include lack of appetite, depression, constipation, diarrhea (which may contain blood), blood in urine, and colic. The exception to livestock and oak toxicity is the domestic pig, which may be fed entirely on acorns in the right conditions, and has traditionally been pastured in oak woodlands (such as the Spanish *dehesa* and the English system of pannage) for hundreds of years.

Acorns are also edible to humans, after leaching of the tannins.

Cultural Significance

Oak branches on the coat of arms of Estonia

National Symbol

The oak is a common symbol of strength and endurance and has been chosen as the national tree of many countries. Already an ancient Germanic symbol (in the form of the Donar Oak, for instance), certainly since the early nineteenth century, it stands for the nation of Germany and oak branches are thus displayed on some German coins, both of the former Deutsche Mark and the current Euro currency. In 2004 the Arbor Day Foundation held a vote for the official National Tree of the United States of America. In November 2004, the United States Congress passed legislation designating the oak as America's National Tree.

Other countries have also designated the oak as their national tree including Bulgaria, Cyprus (Golden Oak), England, Estonia, France, Germany, Moldova, Jordan, Latvia, Lithuania, Poland, Romania, Serbia, and Wales.

Oaks as Regional and State Symbols

The oak is the emblem of County Londonderry in Northern Ireland, as a vast amount of the county was covered in forests of the tree until relatively recently. The name of the county comes from the city of Derry, which originally in Irish was known as *Doire* meaning *oak.*

The Irish County Kildare derives its name from the town of Kildare which originally in Irish was *Cill Dara* meaning the Church of the Oak or Oak Church.

Iowa designated the oak as its official state tree in 1961; and the White Oak is the state tree of Connecticut, Illinois and Maryland. The Northern Red Oak is the provincial tree of Prince Edward Island, as well as the state tree of New Jersey. The Live Oak is the state tree of Georgia, USA.

The oak is a national symbol from the Basque Country, specially in the province of Biscay.

The coat-of-arms of Vest-Agder, Norway, features an oak tree.

Oak leaves are traditionally an important part of German Army regalia. The Nazi party used the traditional German eagle, standing atop of a swastika inside a wreath of oak leaves. It is also known as the Iron Eagle. During the Third Reich of Nazi Germany, oak leaves were used for military valor decoration on the Knights Cross of the Iron Cross. They also symbolize rank in the United States Armed Forces. A gold oak leaf indicates an O-4 (Major or Lt. Commander), whereas a silver oak leaf indicates an O-5 (Lt. Colonel or Commander). Arrangements of oak leaves, acorns and sprigs indicate different branches of the United States Navy Staff corps officers. Oak leaves are embroidered onto the covers (hats) worn by field grade officers and flag officers in the United States armed services.

If a member of the United States Army or Air Force earns multiple awards of the same medal, then instead of wearing a ribbon or medal for each award, he or she wears one metal representation of an "oak leaf cluster" attached to the appropriate ribbon for each subsequent award.

Political Use

The oak tree is used as a symbol by a number of political parties. It is the symbol of Toryism (on account of the Royal Oak) and the Conservative Party in the United Kingdom, and formerly of the Progressive Democrats in Ireland and the Democrats of the Left in Italy. In the cultural arena, the oakleaf is the symbol of the National Trust (UK), The Woodland Trust, and The Royal Oak Foundation.

Religious

Grīdnieku ancient oak in Rumbas parish, Latvia, girth 8.27m, 2015

In Greek mythology, the oak is the tree sacred to Zeus, king of the gods. In Zeus's oracle in Dodona, Epirus, the sacred oak was the centerpiece of the precinct, and the priests would divine the pronouncements of the god by interpreting the rustling of the oak's leaves.

In Baltic and Slavic mythology, the oak is the sacred tree of Latvian Pērkons, Lithuanian Perkūnas, Prussian Perkūns and Slavic Perun, the god of thunder and one of the most important deities in the Baltic and Slavic pantheons.

In Celtic polytheism, the name of the oak tree was part of the Proto-Celtic word for 'druid': **derwo-weyd-* > **druwid-* ; however, Proto-Celtic **derwo-* (and **dru-*) can also be adjectives for 'strong' and 'firm', so Ranko Matasovic interprets that **druwid-* may mean 'strong knowledge'. As in other Indo-European faiths, Taranis, being a thunder god, was associated with the oak tree. The Indo-Europeans worshiped the oak and connected it with a thunder or lightning god; "tree" and *drus* may also be cognate with "Druid," the Celtic priest to whom the oak was sacred. There has even been a study that shows that oaks are more likely to be struck by lightning than any other tree of the same height.

In Norse mythology, the oak was sacred to the thunder god, Thor. Thor's Oak was a sacred tree of the Germanic Chatti tribe.

In the Bible, the oak tree at Shechem is the site where Jacob buries the foreign gods of his people (Gen. 35:4) . In addition, Joshua erects a stone under an oak tree as the first covenant of the Lord (Josh. 24.25–7). In Isaiah 61, the prophet refers to the Israelites as "Oaks of Righteousness." Absalom's long hair (2 Samuel 18:9) gets caught in an oak tree, and allows Joab to kill him.

The badnjak is central tradition in Serbian Orthodox Church Christmas celebration where young and straight oak, is ceremonially felled early on the morning of Christmas Eve.

Historical

Several singular oak trees, such as the Royal Oak in Britain and the Charter Oak in the United States, are of great historical or cultural importance; for a list of important oaks, see Individual oak trees.

"The Proscribed Royalist, 1651", a famous painting by John Everett Millais, depicted a Royalist fleeing from Cromwell's forces and hidden in an oak. Millais painted the picture in Hayes, Kent, from a local oak tree that became known as the Millais Oak.

Approximately 50 km west of Toronto, Canada is the town of Oakville, ON, famous for its history as a shipbuilding port on Lake Ontario.

The city of Raleigh, N.C., is known as "The City of Oaks."

The Jurupa Oak tree – a clonal colony of *Quercus palmeria* or Palmer's oak found in Riverside County, California – is an estimated 13,000 years old.

Large groups of very old oak trees are rare. One of the oldest groups of oak trees, found in Poland, is about 480 years old, which was assessed by dendrochronological methods.

In Republican Rome a crown of oak leaves was given to those who had saved a life of a citizen in battle; it was called the "civic oak".

Famous Oak Trees

Tamme-Lauri oak is the thickest and oldest tree in Estonia.

The Big Oak, by Gustave Courbet (1843).

- The Emancipation Oak is designated one of the 10 Great Trees of the World by the National Geographic Society and is part of the National Historic Landmark district of Hampton University.
- The Ivenack Oak which is one of the largest trees in Europe is located in Mecklenburg-Vorpommern, Germany, and is approximately 800 years old.
- The Bowthorpe Oak, located in Bourne, Lincolnshire, is thought to be 1,000 years old. It was featured in the Guinness Book of World Records and was filmed for a TV documentary for its astonishing longevity.
- The Minchenden (or Chandos) Oak, in Southgate, London, is said to be the largest oak tree in England (already 27 feet or 8.2 meters in girth in the nineteenth century), and is perhaps 800 years old.
- The Seven Sisters Oak is the largest certified southern live oak tree. Located in Mandeville, Louisiana, it is estimated to be up to 1,500 years old with a trunk that measures 38 ft (11.6 meters).
- The Major Oak is an 800- to 1000-year-old tree located in Sherwood Forest, Nottinghamshire. According to folklore, it was used by Robin Hood for shelter.
- Friendship Oak is a 500-year-old southern live oak located in Long Beach, Mississippi.
- The Crouch Oak is believed to have originated in the 11th Century and is located in Addlestone, Surrey. It is an important symbol of the town with many

local businesses adopting its name. It used to mark the boundary of Windsor Great Park. Legend says that Queen Elizabeth I stopped by it and had a picnic.

- The Angel Oak is a southern live oak located in Angel Oak Park on John's Island near Charleston, South Carolina. The Angel Oak is estimated to be in excess of 400–500 years old, stands 66.5 ft (20.3 m) tall, and measures 28 ft (8.5 m) in circumference.
- The Kaiser's Oak, located at the village of Gommecourt in Artois, France, named in honour of Kaiser Wilhelm II, symbolically marked from late 1914 to April 1917 the furthest point in the West of the German Imperial Army during World War One.
- The Wye Oak in Maryland was the United States' largest white oak tree before it blew down in a storm in 2002, at an estimated age of 460 years.

Historical Note on Linnaean Species

Linnaeus described only five species of oak from eastern North America, based on general leaf form. These were white oak, *Quercus alba*; chestnut oak, *Q. montana*; red oak, *Q. rubra*; willow oak *Q. phellos*; and water oak, *Q. nigra*. Because he was dealing with confusing leaf forms, the *Q. montana* and *Q. rubra* specimens actually included mixed foliage of more than one species.

Ruppia

Ruppia, also known as the widgeonweeds, ditch grasses or widgeon grass, is the only genus in the family Ruppiaceae with eight known species). These are aquatic plants widespread over much of the world. The genus name was given in honour of Heinrich Bernhard Rupp, a German botanist (1688-1719). They are widespread outside of frigid zones and the tropics.

Description

The leaf is simple and not rhizomatous. They can be annual (commonly) or perennial (rarely); stem growth is conspicuously sympodial, but sometimes is not. These species are adapted to be in brackish water (and salt marshes). The leaves are small or medium-sized. Their disposition can be alternate, opposite, or whorled (usually alternate except when subtending an inflorescence). Even, lamina keep entire and are setaceous or linear. The leaf just shows one vein without cross-venules. Stomata are not present. The mesophyll leaks calcium oxalate crystals. The minor leaf veins do not present phloem transfer cells and leaks vessels.

These plants have stems without secondary thickening and xylem without vessels. The sieve-tube plastids are P-type. The root xylem does not present vessels.

These plants are hermaphroditic, with anemophilous or hydrophilous pollination. The flowers are ebracteate, small, and regular. Commonly, the flowers are aggregated in 'inflorescences', but sometimes they are solitary. Often, they grow in racemes, spikes, or umbels. The scapiflorous inflorescences are terminal, in short spikes, or subumbelliform racemes, sometimes one- or few-flowered. They do not have hypogynous disks. These flowers do not have perianth absent, except when small staminal appendages are regarded as perianth segments. The androecial members are all equal. The androecium just presents two fertile stamens with sessile anthers dehiscing by longitudinal slits. The pollen is polysiphonous and its grains are three-celled and nonaperturate. The gynoecium (2–)4(–16) is superior, carpelled, and euapocarpous. The carpel is not stylate, apically stigmatic with the stigma peltate, or umbonate. These flowers only present one ovule pendulous, nonarillate, campylotropous, bitegmic, and crassinucellate. The placentation is apical and embryo-sac development is of the polygonum type. Before fertilization, they fuse polar nuclei. The fruit is drupaceous and fleshy, forming an aggregate. The fruiting carpel is indehiscent, commonly on a long, spirally twisted peduncle, with each drupelet becoming very long-stalked. The fruit contains one nonendospermic seed with starch. The embryo can be straight or slightly curved. Membranous testa do not have phytomelan.

Taxonomy

The Cronquist system of 1981 placed the family in order Najadales of subclass Alismatidae in class Liliopsida =monocotyledons in division Magnoliophyta =angiosperms.

The APG II system of 2003 (unchanged from the APG system of 1998) does recognize such a family and places it in the order Alismatales, in the clade monocots.

According to the AP-Website the family is doubtfully distinct from the family Cymodoceaceae: the plants in the three families Cymodoceaceae, Posidoniaceae, and Ruppiaceae form a monophyletic group.

A genus-level taxonomy was briefly revised by Zhao and Wu, including the following species in the world:

species

- *Ruppia bicarpa* - Western Cape, South Africa
- *Ruppia cirrhosa* - temperate regions: Europe, Asia, north + south (but not tropical) Africa, North America, West Indies, Argentina
- *Ruppia didyma* - Mexico, West Indies
- *Ruppia drepanensis* - western + central Mediterranean
- *Ruppia filifolia* - southern South America, Falkland Islands
- *Ruppia maritima* - seashores and lakeshores around the world
- *Ruppia megacarpa* - Australia, New Zealand, Asia (Korea, Japan, and Russia),
- *Ruppia polycarpa* - Australia, New Zealand (incl Chatham Islands)
- *Ruppia tuberosa* - Australia

Marine grasses families: Zosteraceae, Cymodoceaceae, Ruppiaceae and Posidoniaceae. Related families: Potamogetonaceae and (and sometimes including) Zannichelliaceae.

Families and Genera crosses (Sea grasses)			
Kubitzki (ed. 1998)	**Watson & Dallwitz (delta-intkey)**	**data.kew**	**APWeb (mobot.org)**
Zosteraceae			
1. *Zostera* L.	*Zostera*	*Zostera* L.	*Zostera* L. (including *Heterozostera* den Hartog, *Macrozostera* Tomlinson & Posluzny, *Nanozostera* Tomlinson & Posluzny, *Zosterella* J. K. Small)
2. *Heterozostera* den Hartog	*Heterozostera*	*Heterozostera* (Setch.) Hartog	(in *Zostera*)
3. *Phyllospadix* Hook.	*Phyllospadix*	*Phyllospadix* Hook.	*Phyllospadix* J. D. Hooker
Cymodoceaceae			
1. *Syringodium* Kütz	*Syringodium*	*Syringodium* Kutz.	(in *Cymodocea*)
2. *Halodule* Endl.	*Halodule*	*Halodule* Endl.	*Halodule* Endlicher
3. *Cymodocea* König	*Cymodocea*	*Cymodocea* K.Koenig (including *Phycoschoenus* (Asch.) Nakai)	*Cymodocea* König (including *Amphibolis* Agardh ?, *Syringodium* Kütz. ?, *Thalassodendron* den Hartog ?)
4. *Amphibolis* Agardh	*Amphibolis*	*Amphibolis* C.Agardh (including *Pectinella* J.M.Black)	(in *Cymodocea*)
5. *Thalassodendron* de Hartog	(name not found)	*Thalassodendron* Hartog	(in *Cymodocea*)
Ruppiaceae			
Ruppia L.	*Ruppia*	(in *Ruppia* L. in Potamogetonaceae)	*Ruppia* L.
Posidoniaceae			
Posidonia König	*Posidonia*	*Posidonia* K.Koenig	*Posidonia* König

Families and Genera crosses (Potamogetonaceae)

Kubitzki (ed. 1998)	Watson & Dallwitz (delta-intkey)	data.kew	APWeb (mobot. org)
Potamogetonaceae			
1. *Potamogeton* L.	*Potamogeton*	*Potamogeton* L.	*Potamogeton* L. (including *Coleogeton* Les & Haynes, *Stuckenia* Börner)
2. *Groenlandia* J. Gray	*Groenlandia*	*Groenlandia* J.Gay	*Groenlandia* J. Gay
(in *Ruppia* in Ruppiaceae)	(in *Ruppia* in Ruppiaceae)	*Ruppia* L.	(in *Ruppia* in Ruppiaceae)
(in *Althenia*: Zannichelliaceae and *Lepilaena*: Zannichelliaceae)	(in *Althenia*: Zannichelliaceae and *Lepilaena*: Zannichelliaceae)	(in *Althenia* F.Petit: Zannichelliaceae and *Lepilaena* J.L.Drumm. ex Harv.:Zannichelliaceae)	*Althenia* Petit (including *Lepilaena* Harvey)
(in *Pseudalthenia* including *Vleisia*: Zannichelliaceae)	(in *Pseudalthenia*: Zannichelliaceae and *Vleisia*: Zannichelliaceae)	(*Pseudalthenia* not found, *Vleisia* Toml. & Posl.: Zannichelliaceae)	*Pseudalthenia* Nakai (including *Vleisia* Tomlinson & Posluszny)

(in *Zannichellia* L.: Zannichelliaceae)	(in *Zannichellia*: Zannichelliaceae)	(in *Zannichellia* L.: Zannichelliaceae)	*Zannichellia* L.
Zannichelliaceae			
1. *Zannichellia* L.	*Zannichellia*	*Zannichellia* L.	(in *Zannichellia* L.: Potamogetonaceae)
2. *Pseudalthenia* Nakai (including *Vleisia*)	*Pseudalthenia* (excluding *Vleisia*)	(name not found)	(in *Pseudalthenia*: Potamogetonaceae)
3. *Althenia* Petit (excluding *Lepilaena* Drumm. ex. Harv.)	*Althenia* (excluding *Lepilaena*)	*Althenia* F.Petit (excluding *Lepilaena* J.L.Drumm. ex Harv.)	*Althenia* Petit (including *Lepilaena* Harvey)
4. *Lepilaena* Drumm. ex. Harv.	*Lepilaena*	*Lepilaena* J.L.Drumm. ex Harv.	(in *Althenia* Petit)
(in *Pseudalthenia*)	*Vleisia*	*Vleisia* Toml. & Posl.	(in *Pseudalthenia*: Potamogetonaceae)

Phylogeny and Evolution

The first molecular phylogeny of the monogeneric family discerned three distinct species, *R. tuberosa*, *R. megacarpa*, and *R. polycarpa*, and one species complex comprising six lineages. The species complex, named *R. maritima* complex, was later updated as a group of eight lineages. These studies revealed that multiple hybridization and polyploidy events as well as chloroplast capture have occurred in the evolution of the genus.

Heuchera

Heuchera is a genus of evergreen, herbaceous perennial plants in the family Saxifragaceae, all native to North America. Common names include alumroot and coral bells.

Description

Heuchera have palmately lobed leaves on long petioles, and a thick, woody rootstock. The genus was named after Johann Heinrich von Heucher (1677–1746), an 18th-century German physician, and Professor at Wittenberg University. There are approximately 37 species, but the taxonomy of the genus is difficult because the species often intergrade with one another, hybridization is common, and the flowers change markedly in proportion as they develop.

Distribution and Habitat

Alumroot species grow in varied habitats, so some species look quite different from one another, and have varying preferences regarding temperature, soil, and other natural factors. *H. maxima* is found on the Channel Islands of California, where it grows on rocky, windy, saline-washed ocean shores, and *H. sanguinea*, called **coral bells** because of its cerise flowers, can be found in the warm, dry canyons of Mexico and adjacent New Mexico and Arizona. In the Mid-Atlantic region of the United States, the plants grow best in shade.

Several alumroots and their crosses are used as ornamental plants.

Uses

Native American peoples used some *Heuchera* species medicinally. The Tlingit used *H. glabra* as an herbal remedy for inflammation of the testicles caused by syphilis. To the Navajo, *H. novamexicana* was a panacea and a pain reliever. The roots of *H. cylindrica* had a variety of medicinal uses among the Blackfoot, Flathead, Kutenai, Okanagan, Colville, and Shuswap.

Cultivation

The majority of *Heuchera* sold for gardens are hybrids of *H. americana*, such as 'Green Spice'. The original 'Purple Palace' discovered in a palace in England is believed to be a *H. micrantha* × *H. villosa* hybrid, which was then crossed with *H. americana*. Another group of hybrids are crosses of *Heuchera* with *Tiarella* treated under the name × *Heucherella*. Gardeners and horticulturists have developed a multitude of hybrids between various *Heuchera* species. There is an extensive array of blossom sizes, shapes, and colors, foliage types, and geographic tolerances. They are valued as foliage plants, producing rosettes of leaves in shades of green, pink and bronze, often variegated or textured; with long thyrses of white, green, pink or red flowers in spring.

The following cultivars have gained the Royal Horticultural Society's Award of Garden Merit:-

• 'Blackbird'	• 'Fireworks'	• 'Quilter's Joy'
• 'Burgundy Frost'	• 'Magic Wand'	• 'Raspberry Regal'
• 'Can-Can'	• 'Molly Bush'	• 'Sashay'
• 'Chocolate Veil'	• 'Purple Petticoats'	• 'Smokey Rose'

Selected Species

Species include:

- *Heuchera abramsii* - San Gabriel alumroot
- *Heuchera acutifolia*
- *Heuchera alba* - white alumroot
- *Heuchera americana* - American alumroot
- *Heuchera bracteata* - Rocky Mountain alumroot
- *Heuchera brevistaminea* - Laguna Mountain alumroot
- *Heuchera caespitosa* - tufted alumroot
- *Heuchera caroliniana* - Carolina alumroot

- *Heuchera chlorantha* - green-flowered alumroot
- *Heuchera cylindrica* - roundleaf alumroot, lava alumroot
- *Heuchera eastwoodiae* - Senator Mine alumroot
- *Heuchera elegans* - urnflower alumroot
- *Heuchera glabra* - alpine alumroot
- *Heuchera glomerulata* - Chiricahua Mountain alumroot
- *Heuchera grossulariifolia* - gooseberryleaf alumroot
- *Heuchera hallii* - Front Range alumroot
- *Heuchera hirsutissima* - shaggy-haired alumroot
- *Heuchera inconstans*
- *Heuchera lakelae*
- *Heuchera longiflora* - longflower alumroot
- *Heuchera longipetala*
- *Heuchera maxima* - island alumroot, jill-of-the-rocks
- *Heuchera merriamii* - Merriam's alumroot
- *Heuchera mexicana*
- *Heuchera micrantha* - crevice alumroot, smallflower alumroot
- *Heuchera novomexicana* - New Mexico alumroot, range alumroot
- *Heuchera parishii* - Mill Creek alumroot
- *Heuchera parviflora* - littleflower alumroot
- *Heuchera parvifolia* - littleleaf alumroot, common alumroot
- *Heuchera pilosissima* - seaside alumroot
- *Heuchera pubescens* - downy alumroot
- *Heuchera pulchella* - Sandia Mountain alumroot
- *Heuchera richardsonii* - Richardson's alumroot
- *Heuchera rosendahlii*
- *Heuchera rubescens* - pink alumroot, red alumroot, jack-o'the-rocks
- *Heuchera sanguinea* - coral bells
- *Heuchera soltisii*
- *Heuchera villosa* - hairy alumroot, maple-leaved alumroot
- *Heuchera wellsiae*
- *Heuchera woodsiaphila*
- *Heuchera wootonii* - White Mountain alumroot

Hybrids include:

- *Heuchera* × *brizoides*

Gallery

H. micrantha var. *diversifolia*

Heuchera × *brizoides*

H. americana 'Green Spice'

Heuchera 'Starry Night'

REFERENCES

1. Rieseberg. L. H. and Soltis, D. E. (1991) Phylogenetic consequences of cytoplasmic gene flow in plants. Evolutionary Trends in Plants 5: 65-84

2. Terry RG, Nowak RS, Tausch (2000) Genetic variation in chloroplast and nuclear ribosomal DNA in Utah juniper (Juniperus osteosperma, Cupressaceae): evidence for interspecific gene flow. American Journal of Botany 87: 250-258
3. Matos, J. A. and B. A. Schaal. 2000. Chloroplast evolution in the Pinus montezumae complex: a coalescent approach to hybridization. Evolution 54: 1218–1233
4. Whittemore, A. T., Schaal, B. A. (1991) Interspecific gene flow in sympatric oaks. PNAS 88: 2540-2544
5. Ito, Y., T. Ohi-Toma, J. Murata, and Nr. Tanaka (2013) Comprehensive phylogenetic analyses of the Ruppia maritima complex focusing on taxa from the Mediterranean. Journal of Plant Research 126: xxx-xxx.
6. Soltis, D. E., P. S. Soltis, T. G. Collier, and M. L. Edgerton. 1991. Chloroplast DNA variation within and among genera of the Heuchera group: evidence for extensive chloroplast capture and the paraphyly of Heuchera and Mitella. American Journal of Botany 78: 1091–1112.
7. Sunset Western Garden Book, 1995, Leisure Arts, pp. 606–607, ISBN 0376038519.
8. Hogan, C. Michael (2012) Oak Archived 23 May 2013 at the Wayback Machine.. ed. Arthur Dawson. Encyclopedia of Earth. National Council for Science and the Environment. Washington DC
9. Tull, Delena (1999-01-01). Edible and Useful Plants of Texas and the Southwest: A Practical Guide. University of Texas Press. ISBN 9780292781641.
10. Hipp, Andrew (2004). Oak Trees Inside and Out. Rosen Publishing Group, Inc. p. 4.
11. Conrad, Jim. “Oak Flowers”. backyardnature.com. 2011-12-12. Retrieved 2013-11-03.
12. Williams, Joseph H.; Boecklen, William J.; Howard, Daniel J. (2001). “Reproductive processes in two oak (Quercus) contact zones with different levels of hybridisation”. Heredity. 87 (6): 680–690. doi:10.1046/j.1365-2540.2001.00968.x.
13. Arnold, M. L. (1997). Natural Hybridization and Evolution. New York: Oxford University Press. ISBN 0-19-509974-5.
14. Conte, L.; Cotti, C.; Cristofolini, G. (2007). “Molecular evidence for hybrid origin of Quercus crenata Lam. (Fagaceae) from Q-cerris L. and Q-suber L.”. Plant Biosystems. 141 (2): 181–193. doi:10.1080/11263500701401463.
15. Gomory, D.; Schmidtova, J. (2007). “Extent of nuclear genome sharing among white oak species (Quercus L. subgen. Lepidobalanus (Endl.) Oerst.) in Slovakia estimated by allozymes”. Plant Systematics and Evolution. 266 (3–4): 253–264. doi:10.1007/s00606-007-0535-0.
16. Kelleher, C. T.; Hodkinson, T. R.; Douglas, G. C.; Kelly, D. L. (2005). “Species distinction in Irish populations of Quercus petraea and Q. robur: Morphological versus molecular analyses”. Annals of Botany. 96 (7): 1237–1246. doi:10.1093/aob/mci275. PMID 16199484.
17. Frascaria, N.; Maggia, L.; Michaud, M.; Bousquet, J. (1993). “The RBCL Gene

Sequence from Chestnut Indicates a Slow Rate of Evolution in the Fagaceae". Genome. 36 (4): 668–671. doi:10.1139/g93-089. PMID 8405983.

18. Manos, P. S.; Stanford, A. M. (2001). "The historical biogeography of Fagaceae: Tracking the tertiary history of temperate and subtropical forests of the Northern Hemisphere". International Journal of Plant Sciences. 162 (Suppl. 6): S77–S93. doi:10.1086/323280.
19. Raven, Peter H.; Johnson, George B.; Losos, Jonathan B.; Singer, Susan R. (2005). Biology (Seventh ed.). New York: McGraw Hill. ISBN 0-07-111182-4.
20. Baugh, Daniel A. (1965). British Naval Administration in the Age of Walpole. Princeton University Press. p. 242. OCLC 610026758.
21. Cheese. swaledalecheese.co.uk
22. Kala, C.P. (2004). Studies on the indigenous knowledge, practices and traditional uses of forest products by human societies in Uttarakhand state of India. GBPIHED, Almora, India
23. Kala, C.P. (2010). Medicinal Plants of Uttarakhand: Diversity Livelihood and Conservation. BioTech Books, Delhi, ISBN 8176222097.
24. Vohra, D. S. (2004). Bach Flower Remedies: A Comprehensive Study. B. Jain Publishers. p. 3. ISBN 978-81-7021-271-3.
25. "Flower remedies". Cancer Research UK. Retrieved September 2013. Check date values in: |access-date= (help)
26. The Natural Communities of Virginia Classification of Ecological Community Groups (Version 2.3), Virginia Department of Conservation and Recreation, 2010 Archived 15 January 2009 at the Wayback Machine.. Dcr.virginia.gov. Retrieved on 2011-12-10.
27. Schafale, M. P. and A. S. Weakley. 1990. Classification of the natural communities of North Carolina: third approximation. North Carolina Natural Heritage Program, North Carolina Division of Parks and Recreation.
28. "Truffle Glossary: Black Truffles". thenibble.com. 2010-07-01. Retrieved 1 July 2010.
29. "Truffle Glossary: White Truffles". thenibble.com. 2010-07-01. Retrieved 1 July 2010.
30. Kappelle, M. (2006). "Neotropical montane oak forests: overview and outlook", pp 449–467 in: Kappelle, M. (ed.). Ecology and conservation of neotropical montane oak forests. Ecological Studies No. 185. Springer-Verlag, Berlin, doi:10.1007/3-540-28909-7_34 ISBN 978-3-540-28908-1.
31. Lorimer, C.G. (2003) Editorial: The decline of oak forests. American Institute of Biological Sciences.
32. Oldfield, S. & Eastwood, A. (2007) The Red List of Oaks Flora & Fauna International (FFI) and Botanic Gardens Conservation International (BGCI) ISBN 978-1-903703-25-0
33. Kala, C.P. (2012). Biodiversity, communities and climate change. Teri Publications, New Delhi, ISBN 817993442X.

34. Carpenter, Paul (1990). Plants in the Landscape. New York: W.H. Freeman and Company. p. 73. ISBN 0716718081.
35. Waymer, Jim (December 3, 2016). "That's nuts:Acorn onslaught hits Florida". Florida Today. Melbourne, Florida. pp. 1A, 11A. Retrieved December 3, 2016.
36. "Quercus Portal". Retrieved 3 December 2016.
37. Plomion, C., Aury, J.-M., Amselem, J., Alaeitabar, T., Barbe, V., Belser, C., Bergès, H., Bodénès, C., Boudet, N., Boury, C., Canaguier, A., Couloux, A., Da Silva, C., Duplessis, S., Ehrenmann, F., Estrada-Mairey, B., Fouteau, S., Francillonne, N., Gaspin, C., Guichard, C., Klopp, C., Labadie, K., Lalanne, C., Le Clainche, I., Leplé, J.-C., Le Provost, G., Leroy, T., Lesur, I., Martin, F., Mercier, J., Michotey, C., Murat, F., Salin, F., Steinbach, D., Faivre-Rampant, P., Wincker, P., Salse, J., Quesneville, H. and Kremer, A. (2016), Decoding the oak genome: public release of sequence data, assembly, annotation and publication strategies. Mol Ecol Resour, 16: 254–265. doi:10.1111/1755-0998.12425
38. "Trees: Oak Insects and Diseases: Gypsy Moth". TreeHelp.com. Retrieved 27 April 2010.
39. Mougou, A.; Dutech, C.; Desprez-Loustau, M. -L. (2008). "New insights into the identity and origin of the causal agent of oak powdery mildew in Europe". Forest Pathology. 38 (4): 275. doi:10.1111/j.1439-0329.2008.00544.x.
40. Kinver, Mark (28 April 2010). "Oak disease 'threatens landscape'". BBC News. Retrieved 29 April 2010.
41. "Invasion of toxic moths". The Northern Echo. July 10, 2012.
42. Bainbridge, D. A. (12–14 November 1986), Use of acorns for food in California: past, present and future, San Luis Obispo, CA.: Symposium on Multiple-use Management of California's Hardwoods
43. Schierz, Kai Uwe (2004). "Von Bonifatius bis Beuys, oder: Vom Umgang mit heiligen Eichen". In Hardy Eidam; Marina Moritz; Gerd-Rainer Riedel; Kai-Uwe Schierz. Bonifatius: Heidenopfer, Christuskreuz, Eichenkult (in German). Stadtverwaltung Erfurt. pp. 139–45.
44. "Trees – Arbor Day Foundation". Arborday.org. Retrieved 27 April 2010.
45. "Oak Trees". arborday.org. Retrieved 27 April 2010.
46. "Oak as a Symbol". Venables Oak. Retrieved 26 September 2012.
47. "Political or Symbolic". Extended Definition: oak. Retrieved 26 September 2012.
48. Pickles, Eric. "The Conservative Party". Conservatives.com. Retrieved 27 April 2010.
49. Coalition Government 1989 To 1992. progressivedemocrats.ie
50. Frazer, James George (1922). The Golden Bough. Chapter XV: The Worship of the Oak.
51. Taylor, John W. (September 1979). "Tree Worship", Mankind Quarterly, pp. 79–142.
52. Oak. A Dictionary of Literary Symbols. Cambridge.

53. Millais, J.G. (1899) Life and Letters of Sir John Everett Millais, vol. 1, p. 166, London : Methuen.
54. Arborecology, containing a photograph of the Millais oak. arborecology.co.uk
55. Yong, Ed. "The 13,000-year old tree that survives by cloning itself". www.nationalgeographic.com. Retrieved 3 February 2016.
56. Ufnalski K. The oldest groups of oak trees in Poland. Proceedings of EuroDendro 2008 "The long history of wood utilization" News of Forest History Nr. V (39)/2008:83–84
57. Bermosa, Nobert. "Famous Oak Trees in the World". Retrieved 30 September 2012.
58. "Geograph:: Minchenden Oak, Garden of Remembrance,... (C) Christine Matthews". Retrieved 3 December 2016.
59. Seven Sisters Oak. americanforests.org
60. "Seven Sisters Oak". 100 Oaks Project. 4 December 2009.
61. painting by the Swedish botanist C. A. M. Lindman (1856–1928), taken from his book(s) Bilder ur Nordens Flora (first edition published 1901–1905, supplemented edition 1917–1926?)
62. Angiosperm Phylogeny Group (2009). "An update of the Angiosperm Phylogeny Group classification for the orders and families of flowering plants: APG III" (PDF). Botanical Journal of the Linnean Society. 161 (2): 105–121. doi:10.1111/j.1095-8339.2009.00996.x. Retrieved 2013-07-06.
63. Kew World Checklist of Selected Plant Families
64. "Ruppia". Natural Resources Conservation Service PLANTS Database. USDA. Retrieved 26 October 2015.
65. Christenhusz, M. J. M. & Byng, J. W. (2016). "The number of known plants species in the world and its annual increase". Phytotaxa. Magnolia Press. 261 (3): 201–217. doi:10.11646/phytotaxa.261.3.1.
66. Genaust, Helmut (1976). Etymologisches Wörterbuch der botanischen Pflanzennamen ISBN 3-7643-0755-2
67. Zhao L.-C., Wu Z.-Y. (2008) A review on the taxonomy and evolution of Ruppia. Journal of Systematics and Evolution 46: 467–478.
68. Ito, Y., Nr. Tanaka, T. Ohi-Toma, J. Murata, and A.M. Muasya (2015) Phylogeny of Ruppia (Ruppiaceae) revisited: Molecular and morphological evidence for a new species from Western Cape, South Africa. Systematic Botany 40(4): 942-949. doi: 10.1600/036364415X689988
69. Ito, Y., T. Ohi-Toma, J. Murata & Nr. Tanaka (2010) Hybridization and polyploidy of an aquatic plant, Ruppia (Ruppiaceae), inferred from plastid and nuclear DNA phylogenies American Journal of Botany 97: 1156-1167
70. Ito, Y., T. Ohi-Toma, A. V. Skriptsova, M. Sasagawa, Nr. Tanaka, and J. Murata (2014) Ruppia megacarpa (Ruppiaceae): a new species to the floras of Japan, Korea, and Russia. Botanica Pacofica 3: 49–52

71. Kubitzki (ed.) 1998. The families and genera of vascular plants, vol 4, Monocotyledons: Alismatanae and Commelinanae (except Gramineae). Springer-Verlag, Berlin.
72. Watson & Dallwitz. Zosteraceae. The families of flowering plants. http://delta-intkey.com/angio/www/zosterac.htm
73. Vascular Plant Families and Genera. List of genera in family CYMODOCEACEAE (accessed 2016-06-02) http://data.kew.org/cgi-bin/vpfg1992/genlist.pl?CYMODOCEACEAE
74. VASCULAR PLANT FAMILIES and GENERA. List of Genera in CYMODOCEACEAE (accessed 2016-06-02) http://www.mobot.org/mobot/research/apweb/orders/alismatalesweb.htm#Cymodoceaceael
75. Watson & Dallwitz. Potamogetonaceae. The families of flowering plants. http://delta-intkey.com/angio/www/potamoge.htm
76. Vascular Plant Families and Genera. List of genera in family POTAMOGETONACEAE (accessed 2016-06-02) http://data.kew.org/cgi-bin/vpfg1992/genlist.pl?POTAMOGETONACEAE
77. VASCULAR PLANT FAMILIES and GENERA. List of Genera in POTAMOGETONACEAE (accessed 2016-06-02) http://www.mobot.org/mobot/research/apweb/orders/alismatalesweb.htm#Potamogetonaceae
78. Ito, Y., T. Ohi-Toma, J. Murata & Nr. Tanaka (2013) Comprehensive phylogenetic analyses of the Ruppia maritima complex focusing on taxa from the Mediterranean Journal of Plant Research 126: 753-762
79. Sunset Western Garden Book. 1995. 606–607.
80. Heuchera
81. Phipps, Nikki. "Planting Coral Bells: Tips For Growing The Coral Bells Plant In Your Garden". gardeningknowhow.com. Retrieved 21 December 2016.
82. Heuchera. Flora of North America.
83. Archibald William Smith A Gardener's Handbook of Plant Names: Their Meanings and Origins , p. 185, at Google Books
84. Coombs, George (2014). "Heuchera for the Mid-Atlantic Region". Mt. Cuba Center. Retrieved 27 January 2017.
85. Heuchera glabra. Flora of North America.
86. Heuchera novamexicana. Flora of North America.
87. Heuchera cylindrica. Flora of North America.
88. Armitage's Garden Perennials. Retrieved 2013-04-01 – via Google Books.
89. Cullina, W. New England Wild Flower Society Guide to Growing and Propagating Wildflowers of the United States and Canada. Houghton Mifflin Harcourt, 2000. ISBN 0-395-96609-4, ISBN 978-0-395-96609-9
90. "RHS Plant Selector Heuchera 'Blackbird' / RHS Gardening". Apps.rhs.org.uk. Retrieved 2013-04-01.

91. "RHS Plant Selector Heuchera 'Burgundy Frost' / RHS Gardening". Apps.rhs.org.uk. Retrieved 2013-04-01.

92. "RHS Plant Selector Heuchera 'Can-can' AGM / RHS Gardening". Apps.rhs.org.uk. Retrieved 2013-04-01.

93. "RHS Plant Selector Heuchera 'Chocolate Veil' / RHS Gardening". Apps.rhs.org.uk. Retrieved 2013-04-01.

94. "RHS Plant Selector Heuchera 'Fireworks' PBR AGM / RHS Gardening". Apps.rhs.org.uk. Retrieved 2013-04-01.

95. "RHS Plant Selector Heuchera 'Magic Wand' AGM / RHS Gardening". Apps.rhs.org.uk. Retrieved 2013-04-01.

96. "RHS Plant Selector Heuchera 'Molly Bush' AGM / RHS Gardening". Apps.rhs.org.uk. Retrieved 2013-04-01.

97. "RHS Plant Selector Heuchera 'Purple Petticoats' AGM / RHS Gardening". Apps.rhs.org.uk. Retrieved 2013-04-01.

98. "RHS Plant Selector Heuchera 'Quilter's Joy' / RHS Gardening". Apps.rhs.org.uk. Retrieved 2013-04-01.

99. "RHS Plant Selector Heuchera 'Raspberry Regal' AGM / RHS Gardening". Apps.rhs.org.uk. Retrieved 2013-04-01.

100. "RHS Plant Selector Heuchera 'Sashay' AGM / RHS Gardening". Apps.rhs.org.uk. Retrieved 2013-04-01.

101. "RHS Plant Selector Heuchera 'Smokey Rose' / RHS Gardening". Apps.rhs.org.uk. Retrieved 2013-04-01.

102. Heuchera: List of Species. Flora of North America.

103. Heuchera. USDA PLANTS.

104. GRIN Species Records of Heuchera. Germplasm Resources Information Network (GRIN).

Index